T0269582

CAMBRIDGE LIBRARY COLLECTION

Books of enduring scholarly value

Technology

The focus of this series is engineering, broadly construed. It covers technological innovation from a range of periods and cultures, but centres on the technological achievements of the industrial era in the West, particularly in the nineteenth century, as understood by their contemporaries. Infrastructure is one major focus, covering the building of railways and canals, bridges and tunnels, land drainage, the laying of submarine cables, and the construction of docks and lighthouses. Other key topics include developments in industrial and manufacturing fields such as mining technology, the production of iron and steel, the use of steam power, and chemical processes such as photography and textile dyes.

The Life of Robert Stephenson, F.R.S.

Relying on incremental experiment and practice rather than individual leaps into the unknown, Robert Stephenson (1803–59) forged an influential career as a highly respected railway and civil engineer. From the steam locomotive *Rocket* to the London and Birmingham Railway and the Britannia Bridge, his work helped to consolidate the foundations of the modern engineering profession. Based on the first-hand testimony of relatives and contemporaries as well as correspondence and official records, this 1864 biography by John Cordy Jeaffreson (1831–1901), published only five years after Stephenson's death, tells the story of this quiet industrial innovator. Five chapters by engineer William Pole (1814–1900) provide a more technical insight, examining some of Stephenson's most significant railway bridges and his involvement with the atmospheric system. Volume 1 traces Robert's early life, carefully moulded by his father George, and also covers the building of the London and Birmingham Railway.

The Life of
Robert Stephenson, F.R.S.

*With Descriptive Chapters on Some
of his Most Important Professional Works
by William Pole*

VOLUME 1

JOHN CORDY JEAFFRESON

CAMBRIDGE
UNIVERSITY PRESS

CAMBRIDGE
UNIVERSITY PRESS

University Printing House, Cambridge, CB2 8BS, United Kingdom

Published in the United States of America by Cambridge University Press, New York

Cambridge University Press is part of the University of Cambridge.
It furthers the University's mission by disseminating knowledge in the pursuit of
education, learning and research at the highest international levels of excellence.

www.cambridge.org
Information on this title: www.cambridge.org/9781108070744

© in this compilation Cambridge University Press 2014

This edition first published 1864
This digitally printed version 2014

ISBN 978-1-108-07074-4 Paperback

THE LIFE

OF

ROBERT STEPHENSON.

VOL. I.

LONDON
PRINTED BY SPOTTISWOODE AND CO.
NEW-STREET SQUARE

George Richmond 1849   Henry Adlard 1863

London Longman & Co.

THE LIFE

OF

ROBERT STEPHENSON, F.R.S.

ETC. ETC.

LATE PRESIDENT OF THE INSTITUTION OF CIVIL ENGINEERS.

BY

J. C. JEAFFRESON

BARRISTER-AT-LAW.

WITH DESCRIPTIVE CHAPTERS ON

SOME OF HIS MOST IMPORTANT PROFESSIONAL WORKS

BY

WILLIAM POLE, F.R.S.

MEMBER OF THE INSTITUTION OF CIVIL ENGINEERS.

IN TWO VOLUMES.

VOL. I.

LONDON:

LONGMAN, GREEN, LONGMAN, ROBERTS, & GREEN.

1864.

THE LIFE

OF

ROBERT STEPHENSON, F.R.S.

PREFACE.

FOUR YEARS have elapsed since with Professor Pole I undertook to write the Life of ROBERT STEPHENSON.

A careful examination of the many published works which, either specially or incidentally, treat of the labours of the two Stephensons, was amongst the first steps which I took towards the performance of my task. I read critically a large number of scientific volumes, biographies, lectures, and articles bearing upon the history of the locomotive, upon the art of building bridges, and upon the careers of the men who, during the last sixty years, have brought our railway system to its present state of efficiency. My surprise was great at finding that the statements of the various treatises were irreconcilable.

In the summer and autumn of 1860 I passed some time in Northumberland and Durham, collecting materials for this work from the oral communications of Robert Stephenson's numerous relations, from the reminiscences of men who had been the companions or the patrons of both the Stephensons, and from entries in parish registers, and the account-books of collieries and factories. I was fortunate in meeting with cordial response from all of the many persons whose assistance was solicited. The result of these enquiries was the discovery that many mistakes had been made in telling the story of the elder Stephenson's life, and that no life of the younger Stephenson would be complete that should neglect to give a correct account of the misapprehended passages in the life of the elder. The only course, therefore, open to me was to re-write the Life of George Stephenson, so far as it affected Robert Stephenson's career, and to tell the whole truth of the son's life to the best of my ability.

On my return from the North of England I gathered documentary materials from many different quarters, and ere long I was fortunate enough to bring together a mass of evidence which the representatives of Robert Stephenson did not know to be in existence. Besides letters submitted to my perusal by a great number of the engineer's friends, and besides papers sent to me by his executors, I obtained custody of several important collections of documents. Mr. Longridge put into my hands the Stephenson papers which his father preserved. Mr. Illingworth allowed me to peruse his South-American papers. Mr. Charles Empson, shortly before his death, contributed to my store of materials a most interesting collection of letters and documents; consisting of Robert Stephenson's early journals, and of nearly all the letters which he either received *from* or had written *to* friends or relations, between the termination of his life on Killingworth Moor and his return from South America. I have also to acknowledge the assistance of Mr. George Parker Bidder, late President of the Institution of Civil Engineers; Mr. Charles Manby, F.R.S.; and Mr. George Robert Stephenson, C.E.

In expressing my thanks to the gentlemen who have assisted me with information or papers, I render no mere formal act of courtesy. Gratitude is a solemn duty when acknowledgment has to be made of services conferred by those who no longer tarry in the ways of men. Of those to whom I am indebted for facts or counsel, many have passed to another world. Mr. Losh and Mr. Weallens of Newcastle, Mr. Kell of Gateshead, Mr. Charles Empson of Bath, Admiral Moorsom, and Mr. Charles Parker, are amongst those who will never see this page.

J. CORDY JEAFFRESON.

THE TASK of describing some of the more important professional subjects which occupied the attention of Robert Stephenson has been confided to me. There was some difficulty in determining what subjects should be chosen, for many of his works were so mixed up with the current events of his life, that they could scarcely be separated from the narrative of his biography.

I determined, finally, to select the Atmospheric system of Railway Propulsion, and the great Iron Railway Bridges erected by him.

The length at which I have treated the former of these subjects demands some explanation, inasmuch as Robert Stephenson, far from promoting the Atmospheric system, was always one of its strongest opponents. But judges on whom I can fully rely were of opinion that it deserved a prominent place in his life, as well from the great interest he took in it, as from the extent to which it must have affected the whole course of Railway engineering. The facts of its history, with the results and lessons to be drawn from it, seemed likely soon to be forgotten, and were considered worthy of being put fully on record.

The preliminary chapter on Iron Bridges has been written in order to bring out more clearly the peculiarities and merits of the magnificent structures of this kind, to which probably Robert Stephenson will eventually owe his widest fame.

I have to acknowledge information kindly supplied by many friends in the profession.

The chapters which I have contributed to the work are XIV. in Vol. I., and II., III., IV., VIII., in Vol. II.

WILLIAM POLE.

LONDON: *September* 1864.

CONTENTS

OF

THE FIRST VOLUME.

———◇———

CHAPTER VI.

SOUTH AMERICA.

(ÆTAT. 20-24.)

CHAPTER VII.

FROM SOUTH AMERICA TO NEWCASTLE.

(ÆTAT. 23-24.)

CHAPTER VIII.

RESIDENCE IN NEWCASTLE.

(ÆTAT. 24-25.)

CHAPTER IX.

RESIDENCE IN NEWCASTLE—CONTINUED.

(ÆTAT. 25-28.)

CHAPTER X.

CONSTRUCTION OF THE LONDON AND BIRMINGHAM RAILWAY.

(ÆTAT. 29-34.)

CHAPTER XI.

AFFAIRS, PUBLIC AND PRIVATE, DURING THE CONSTRUCTION OF THE LONDON AND BIRMINGHAM RAILWAY.

(ÆTAT. 29-35.)

CHAPTER XII.

FROM THE COMPLETION OF THE LONDON AND BIRMINGHAM RAILWAY TO THE OPENING OF THE NEWCASTLE AND DARLINGTON LINE.

(ÆTAT. 35-41.)

CHAPTER XIII.

RAILWAY PROGRESS AND RAILWAY LEGISLATION.

CHAPTER XIV.

THE ATMOSPHERIC SYSTEM OF RAILWAY PROPULSION.

ILLUSTRATION IN VOL. I.

THE LIFE

OF

ROBERT STEPHENSON.

———o૦ૐ૦o———

CHAPTER I.

THE STEPHENSON FAMILY.

Various Stephensons of Newcastle — 'Old Robert Stephenson' — Mabel Carr — George Stephenson's Birth — Fanny Henderson — George Stephenson moves to Willington — Robert Stephenson's Birth — The Christening Party at Willington Quay — Mrs. George Stephenson's delicate Health — George Stephenson removes to Killingworth Township, Long Benton — Site of George Stephenson's House at Willington — 'The Stephenson Memorial.'

THE records of Newcastle show that the name of Stephenson has been frequent in every rank of the town for the last two hundred and fifty years. But no attempt has ever been made to establish a family connection between the subject of this memoir and the many worthy citizens of Newcastle who, in the seventeenth and eighteenth centuries, bore the same name. A gentleman of high attainments, residing in the neighbourhood of Newcastle, in answer to enquiries for ancestors in the male line of George Stephenson, stated that George Stephenson on a certain occasion said that his family were natives of Castleton, in Liddisdale, and that his grandfather came into England in the service of a Scotch gentleman.

There is no doubt that the grandfather of the greatest engineer of the present century lived and toiled and died in humble circumstances. He worked as fireman to the engines of the various colliery pits in the neighbourhood of Wylam, till an accident deprived him of sight and rendered him dependent on others for his daily bread. Gentle beyond the wont of rude North-countrymen, and fond of spinning out long stories of adventure and romance to village children, he was known as 'Bob the story-teller.' He is now remembered by the few of his associates who linger on the earth as 'Old Robert Stephenson.' In early life he married Mabel, the daughter of George Carr, a bleacher and dyer of Ovingham, a village standing on an ascent which rises from the north bank of the Tyne, and faces the ancient ruins of Prudhoe Castle, that crown the hill on the opposite bank. The maiden name of Mabel Carr's mother was Eleanor Wilson. Eleanor was the daughter of a wealthy Northumbrian yeoman, who possessed a good estate in the parishes of Stocksfield and Bywell. Indignant at her marriage with the bleacher and dyer of Ovingham, Mr. Wilson turned his back upon her, and died without bequeathing her a penny.

By his wife Mabel 'Old Robert Stephenson' had four sons (James, George, Robert, and John) and two daughters (Eleanor and Ann). James, the eldest son, closely resembled his father; but George, Robert, and

John, were all shrewd and observant men, self-reliant and resolute.

Born June 9, 1781, George Stephenson could neither write nor read when he had attained the age of eighteen years. Up to that age he displayed no signs of unusual intelligence, but he had always been a good, sober, steady lad. Like most pit-children, he used to grub about in the dirt, and for his amusement fashion models of steam-engines in clay. From his earliest years, also, he kept as pets pigeons, blackbirds, guinea-pigs, and rabbits ; an almost universal trait amongst the colliery labourers of the Newcastle field.

In 1801, he became brakesman of the engine of the Dolly Pit, in Black Callerton, and lodged in the house of Thomas Thompson, a small farmer of that parish. George Stephenson was at that time a light-hearted young fellow, famous for practical jokes, and proud of his muscular power. At this period, also, he acquired the art of shoe-cobbling.

The most important farmer of the parish was Mr. Thomas Hindmarsh, who occupied land which his an-cestors had farmed for at least two centuries. To his grave displeasure, his daughter Elizabeth accepted the addresses of the young brakesman, giving him clandestine meetings in the orchard and behind the garden-fence, until such effectual measures were taken as prevented a repetition of the suitor's visits. Elizabeth, however, remained faith-ful to the lover, whom her father drove from his premises, and she eventually became his second, but not his last, wife.

George Stephenson took this disappointment lightly. He soon fixed his affections on Ann Henderson, daughter of John Henderson, a small and impoverished farmer, near Capheaton. Like her two sisters Hannah and Frances (who were the female servants in Thomas Thompson's house) Ann was a domestic servant. At first she seemed well pleased with her lover, who, amongst other attentions, paid her one which deserves a few passing words.* Observing that her shoes wanted to be re-soled, he begged leave to mend them, and, the permission being granted, he not only repaired them, but boastfully displayed them to his companions. His triumph, however, was of short duration; for on returning the shoes to Ann, with a request for a warmer acknowledgement of his services than mere thanks, he was informed by her that he wooed where he could never win.

This second rejection was for a time deeply felt, but he concealed his chagrin, and then made up his mind that, since he could not have Ann, he would try his luck with her sister Fanny.

Fanny Henderson had for years been a servant in the house where George Stephenson was a lodger. When Thomas Thompson, more than ten years before, took the farm from the outgoing tenant, George Alder, she came into his service as part of the concern, with the following character : —

---

* Mr. Pattison, the nephew of Ann Henderson, and son of Elizabeth Henderson (who married Thomas Pattison, a tenant farmer of Black Callerton), writes thus : ' The pair of shoes mentioned in the " Life of George Stephenson," as having been made for Fanny Henderson, afterwards his wife, were not for her, but for her sister Ann, whom he ardently admired ; but not succeeding with her, he said he would have one of the family, and he turned his attention to Fanny.' Mr. Pattison, the author of this statement, is employed in the factory of Messrs. Robert Stephenson and Co., Newcastle. His statement is corroborated by all the members of his mother's respectable family.

Black Callerton: April 10, 1791.

The bearer, Frances Henderson, is a girl of a sober disposition, an honest servant, and of a good family, as witness my hand,

GEORGE ALDER.

She was no longer young, and it was the village gossip that she would never find a husband. As a girl, she had plighted her troth to John Charlton, the village school-master of Black Callerton, but their long engagement was terminated in 1794 by the young man's death, when she was in her twenty-sixth year. She was therefore George's senior by twelve years; but it was not for her to object to the disparity of their ages, since he was willing to marry a woman so much older than himself. So, to the good-natured amusement of neighbours, and to the vexation of Ann Henderson, who did not enjoy the apparent unconcern with which her lover had passed from her to her old maid sister, George Stephenson was married at Newburn church on November 28th, 1802, to Fanny Henderson, the mother of the subject of this memoir.

Mr. Thomas Thompson gave the wedding breakfast to his faithful domestic servant and his young lodger, and signed his name in the parish register, as a witness of the marriage ceremony. George Stephenson had at that time so far advanced in the art of writing, that he was able to sign his own name (and his wife's maiden name also — if handwriting may be trusted as evidence on such a point) on the certificate. The signature is blurred — possibly by the sleeve of his coat, as he stretched out his pen for another dip of ink before acting as his wife's secretary; but the handwriting is legible, and is a good specimen of George Stephenson's caligraphy.

For a short time after his marriage George Stephenson continued to reside at Black Callerton, lodging with his wife in a cottage not far from the Lough House, as Mr. Thomas Thompson's residence was called. This arrangement, however, did not last long. While he was acting as brakesman at Black Callerton, his father and his brothers James and John continued to work at Walbottle colliery, where the engineer was Robert Hawthorn, the ingenious and enterprising man whose sons still carry on the important locomotive factory at Newcastle that bears their name. At the opening of the present century, Robert Hawthorn, then known as one of the best enginewrights in the Newcastle country, erected the first ballast machine that ever worked on the banks of the Tyne. This machine was erected at Willington Quay (a station on the river side, about six miles below Newcastle), and was placed upon the quay, on the edge of the river.* When the work was completed, Hawthorn exerted his influence in favour of the Dolly Pit brakesman, the consequence of which was, that the latter quitted Black Callerton (situated a few miles above Newcastle), and became the brakesman of the Ballast Hill engine. It was while he held this appointment that George Stephenson first set up as a housekeeper on his own humble account — that is to say, first bought bedding and such modest furniture as he required for

---

* It has been represented that this machine was placed on the summit of the Ballast Hill. The Messrs. Hawthorn, however (the sons of the contractor), who remember well both the engine and the incline, say that the former was near the water. 'If,' say these gentlemen, 'the machine had been erected on the open Ballast Hill, it would have been buried up.'

two rooms in a cottage stationed hard by the engine on Willington Quay. As everything connected with the career of this remarkable man is interesting, it is worthy of mention that at the time of his marriage he had not saved sufficient money to buy the upholstery and fittings of his new home. In marrying Fanny Henderson, however, he had, in a pecuniary sense, bettered himself. When they mounted the horse which Mr. Burn of the Red House farm, Wolsingham, put at their service, and made their progress from their furnished lodgings at Black Callerton to their new domicile on the other side of Newcastle, George had in his pocket a handsome number of gold pieces — the savings of his careful wife during long years of domestic service. A portion of this money was expended on household goods, the rest being laid by against a rainy day.

Marriage made a great difference in George Stephenson, and on settling at Willington he applied himself earnestly to the work of self-education. On October 16th, 1803,* his wife gave birth to a son, who was christened Robert: the ceremony was performed in the Wallsend school-house, as the parish church was unfit for use. The sponsors were Robert Gray and Ann Henderson, but they were by no means the only guests at the christening. Proud of being a father, George called together his kinsmen from the Wylam and Newburn districts, and gave them hospitable entertainment. His father, mother, and brothers answered the summons. So Robert Stephenson was received into the family with

---

* Robert Stephenson stated that he was born in the month of November, and his birthday was always celebrated at that time; but the extract from the register proves his birth to have been in October.

all honour, being named, according to north-country
fashion, after his grandfather, and having long life and
health and success drunk to him in sound ale and
Scotch whisky.   But the uncles and aunts who were
present at the festivities remarked that the babe was
' a wee sickly bairn not made for long on this earth.'

Delicate the child both was and remained until he had
made several years' entry into manhood.   From his father
he inherited strong thews and a strong will; but from
his mother's blood there was a taint imparted to what
otherwise would have been a magnificent constitution.
The disease — consumption — which carried off John
Charlton, now made insidious advances on Mrs. Stephenson; and her husband, whilst he was still only two and
twenty years old, saw his life darkened by the heaviest
misfortune that can befall a poor man — an invalid wife.
In this respect his career sadly resembled the lot of his
father, and years afterwards it was mournfully reproduced
in the experiences of his only son.

But the young father was not the man to crouch
at the first blast of adversity.   If his wife could not
help him, the more reason that he should help himself.
He worked steadily at his engine during the appointed
hours, and employed his evenings in shoemaking and
cobbling and in acquiring the rudiments of mechanics.
Whilst he was spelling out the secrets of his books, and
often as he worked, hammer in hand, he relieved his
sickly wife by taking his son from her cough-racked
breast and nursing him for hours together.   Robert's
earliest recollections were of sitting on his father's knee,
watching his brows knit over the difficult points of a
page, or marking the deftness and precision with which

his right hand plied its craft. The child, too, bore in body as well as heart a memorial of his father's tenderness. His seat was always on George's left knee, his body encircled by his father's left arm. The consequence was that the left hand and arm, left at liberty by the position, became stronger and were more often used than the right; and the child's habit of trusting the left hand, strengthening with time, gradually developed into a permanent defect.

George Stephenson did not remain long at Willington, but his brief residence on the quay side was marked by other incidents besides the birth of his child. It was there that his intercourse with Robert Hawthorn first took the form of personal intimacy. It was at Willington, too, that he first took to clock-mending and clock-cleaning as an additional field of industry. The pit-man's cabin has points by which it may be distinguished from the southern peasant's cottage. Its prominent article of furniture is a good and handsome bed. Not seldom a colliery workman spends ten, or fifteen pounds on his bedstead alone, and when he has bought the costliest he can afford he places it in the middle of his principal apartment. Invariably he has also a clock—usually a valuable one—amongst his possessions. Every village, therefore, abounds in clocks, and as the people are very particular and even fanciful about them, a brisk business is everywhere carried on by clock-cleaners. Each petty district has its own clock-cleaner, who is supported by all the inhabitants; and it is to be observed that this artificer almost invariably has been self-taught.

George Stephenson, therefore, in occupying his spare

time in cleaning clocks, did only what the superior and more intelligent workmen of his time and country were in the habit of doing. His new employment was lucrative, and enabled him, for the first time in his life, to lay by money out of his own earnings.

Recent circumstances have connected the Stephensons in the public mind with Willington; but their relations with that township were neither lasting nor intimate. Scarcely had George Stephenson formed attachments to his neighbours when he moved to the parish of Long Benton, where he was engaged as brakesman of the West Moor colliery engine. On receiving his new appointment, George, now twenty-three years old, with his wife and little Robert (then in his second year), settled in a cottage in Killingworth township, close to the West Moor colliery—about four or five miles to the north of Newcastle, and about the same distance from Willington Quay.

The cottage in which George Stephenson lived on Willington Quay has been pulled down, but before it was destroyed the public interest attaching to it was so great, that photographic pictures and engravings of it had been circulated in every direction. The site, however, of Robert Stephenson's birthplace is appropriately preserved. Of the objects which arrest the attention of a person making the passage up the river from Tynemouth to Newcastle, there is nothing of greater architectural merit than the Gothic edifice that stands out upon Willington Quay. This structure, generally spoken of as the 'Stephenson Memorial,' comprises (besides rooms for officers and teachers) two school-rooms, one for boys and another for girls, and a reading-room for mechanics. The entire building is a model of what such a structure ought to be, and the children's play-grounds are as spacious and well-appointed as the interior of their excellent institution. The exact spot on which the Stephenson cottage stood, is now the boys' play-ground, in the rear of the school.

CHAPTER II.

LONG BENTON.

(ᴂᴛᴀᴛ. 1-9.)

The West Moor Colliery —'The Street' of Long Benton — Road from Newcastle to Killingworth —' The Cottage' on the West Moor — View from the Cottage Windows—Apparent Amendment of Mrs. Stephenson's Health — Robert and his Mother visit Black Callerton — Robert Stephenson's Sister — Death of his Mother — George Stephenson's Journey to Montrose — Eleanor Stephenson — Her great Disappointment — 'The Artificials' — Little Robert's Visits to the Red House Farm, Wolsingham — ' The Hempy Lad ' — Tommy Rutter's School — The young Gleaner — A Lesson for the Lord's Day — George Stephenson's Sundays — His Friends, Robert Hawthorn and John Steele—The first Locomotive ever built on the Banks of the Tyne — Anthony Wigham —Captain Robson—Evenings at the West Moor.

TOWARDS the close of 1804, George Stephenson moved to the West Moor colliery, and fixed himself and family in the little cottage where he resided, till he made rapid strides to opulence and fame. Long Benton,* a wide straggling parish, comprising in its five townships numerous colonies of operatives, presents those contrasts of wealth and poverty for which mining and manufacturing districts are proverbial. The long irregular street of the village is not without beauty. The vicarage

---

* In this parish Smeaton, in 1772, erected the large atmospheric engine, which formed the standard engine before Watt's improvements.—W. P.

is a picturesque dwelling, and on either side of the road, surrounded by gardens, with paths of crushed slag and refuse coal, and plantations of a somewhat sooty hue, are the houses of prosperous agents and employers. The general aspect of the place, however, is humble, and the abodes of the poorer inhabitants are comfortless.

The road from Newcastle to Long Benton quits the town at the northern outskirt, and, leaving ' the moor ' on the left, passes through the picturesque plantations of Jesmond Vale (watered by the brawling Dean that flows to Ouseburn), and, having ascended the bold and richly wooded sweep of Benton Banks, leads on over a bleak and unattractive level to Long Benton, where art and nature again combine to render the landscape attractive. Pursuing its course down the disjointed village, the road descends to the church, where it turns to the left over a rustic stone bridge, curves round a corner of the churchyard, and bears away to Killingworth township and the West Moor colliery.

The cottage in which the young brakesman and his middle-aged wife settled, was a small two-roomed tenement. Even as it now stands, enlarged by George Stephenson to the dignity of a house with four apartments, it is a quaint little den — a toy-house rather than a habitation for a family. The upper rooms are very low, and one of them is merely a closet. The space of the lower floor is made the most of, and is divided into a vestibule and two apartments. Over the little entrance door, in the outer wall, is a sun-dial, of which mention will be made hereafter. The principal room of the house is on the left hand of the entrance, and in it stands to this day a piece of furniture which is now the property of Mr. Lancelot Gibson, the hospitable occupant of the cottage. This article of furniture is a high strong-built cheffonier, with a book-case surmounting it, and it was placed in the apartment by George Stephenson himself. Of this chattel mention will be made elsewhere in these pages.

The view from the little garden, in front of this cabin, is as fine as any in the neighbourhood of Newcastle. A roadway leading to the North Shields turnpike road runs along the garden rails; on the other side of the road is a small paddock, not a hundred yards in width, beyond the farther confine of which are the mud walls of the glebe farmhouse, of which George Stephenson's friend Wigham was tenant. On the right hand, buried in trees, is Gosforth Hall, formerly the residence of the Mr. Brandling who fought George's battle in the matter of the safety lamp, and whose name — though he has long been dead — is never mentioned by the inhabitants of the district without some expression of affectionate regard. Newcastle cannot be seen; but clearly visible is the blue-hill ridge beyond it, on the farther decline of which rests the seat of the Liddells—Ravensworth Castle.

The excitement of moving to Killingworth was for a time beneficial to Mrs. Stephenson's health. She became more cheerful; and, that she might have every chance of amendment, George Stephenson prevailed on her to visit her sister Elizabeth, who had married Thomas Pattison, a farmer of Black Callerton.

This apparent improvement in health, which her husband attributed altogether to the excitement of moving to a new home, was, however, little more than the ordinary consequence of pregnancy, which is well known to stay for a brief space the treacherous incursions of phthisical malady. In the July of 1805 she was put to bed, and Robert Stephenson had a sister who lived just three weeks *— long enough to be named Frances after her mother, to be admitted into Christ's Church, and to taste something of human suffering. Her little girl born, dead, and buried, the bereaved mother relapsed into her previous condition. The cold winter and spring, with its keen north-eastern winds sweeping over the country, completed the slow work of consumption, and before Benton banks and Jesmond vale had again put forth their green leaves, she was quiet in her last earthly rest in Benton churchyard.

Deprived of his mother, before he had completed his third year, Robert Stephenson was placed under the care of the women who were successively George's housekeepers. Of the three housekeepers who lived in the West Moor cabin, the first and last were superior women. Soon after the death of his wife, George Stephenson went for a few months to Scotland, where he was employed as engineer in a large factory near Montrose. On making this journey, he left little Robert in the custody of his first housekeeper, at Killingworth. On his return he was surprised, and slightly angry, at finding his house shut up, and without inmates. In his absence, the housekeeper (who was in every respect an excellent woman) had become the wife of his

---

* The Long Benton registers contain the following entries: —

1. Frances Stephenson, West Moor Colliery, d. of George Stephenson and Frances his wife, late Henderson. Died Aug. 3, 1805. Buried Aug. 4.

Aged 3 weeks.

2. Buried 1806, Frances Stephenson, late Henderson, West Moor, wife of George Stevenson (sic). Died May 14. Buried May 16. Aged 37 years.

brother Robert, in whose dwelling the little boy then was. Recovering possession of his child, George Stephenson again established himself at the West Moor, engaged a second housekeeper, and, having well-nigh emptied his pockets by paying some debts of his poor blind father, and by purchasing a substitute for service in the militia, once more set to work resolutely as brakesman, cobbler, and clock-cleaner. The burden of an invalid wife, of which he had been relieved, was replaced by the burden of a helpless father. Struck blind by an accident which has been already mentioned, ' Old Robert ' was maintained in comfort by his sons until the time of his death.

George's second selection of a housekeeper was not so fortunate as his first, but he soon dismissed her, and received into his cottage his sister Eleanor, or, as her name is spelt in the family register, Elender. This worthy and pious woman, born on April 16, 1784, was nearly three years the junior of her brother, and consequently was still young when she came to keep his house. But young as she was, she had made acquaintance with sorrow. A merry lass, she went up to London to fill a place of domestic service, having first plighted her troth to a young man in her own rank of life, under a promise to return and become his bride whenever he wished to marry her. A year or two passed, when, in accordance with this agreement, her lover summoned her back to Northumberland. Eleanor went on board a Newcastle vessel homeward bound. Ill-fortune sent adverse breezes. The passage from the Thames to the Tyne consumed three weeks, and when the poor girl placed her foot on the quay side of the Northumbrian capital, the first piece of intelligence she received was that her faithless lover was already the husband of another.

George Stephenson invited his sister to his house, and she, seeing a field of usefulness before her, wisely accepted the invitation. Her sister Ann having already married, and migrated to the United States, Eleanor was to George as an only sister.

The record of one trifling but pathetic difference between George and Eleanor is still preserved by family gossip. When Eleanor first took up her abode at the West Moor colliery, she wore some cheap artificial flowers in her bonnet. The sad experiences of the four preceding years had made the young brakesman less gentle in his temper and more practical in his views. Rude love of truth and dislike of *shams* caused him to conceive a dislike for these 'artificials,' as he contemptuously termed them. He asked Eleanor to throw them away, but she, averring that they cost good money, declined to do so.

'Nay, then,' said George, stretching out his hand, 'let me take them out and throw them away, and I'll give thee a shilling.'

But Eleanor, usually so meek and gentle, drew back. George saw her secret and blundered out an apology. The poor girl had put those flowers in her bonnet, in the vain hope that they would render her comely face more acceptable to her false lover. She had been rightly punished for what she called her worldly vanity ; and in humble acknowledgment of her error, she determined to wear ' the artificials ' as a memorial of her foolishness.

From her early days she had been seriously inclined ;

and her recent disappointment gave a tone to her mind
that was not to be outgrown. Joining the Wesleyan
Methodists, she regularly attended their prayer-meetings;
and all who remember her bear witness that her labours
of unassuming charity aptly enforced the teaching of her
lips. Her spare hours were employed in visiting the
sick, and repeating long passages from the Bible to those
who were themselves unable to spell out the secrets of
' the Word.'

It was a bright day for little Robert when this young
woman entered the cottage at the West Moor, and took
him into her affectionate keeping. The best and most
pleasant glimpses that can be obtained of his childhood,
show the healthiest relations to have subsisted between
him and this good aunt.

Every few months Aunt Nelly used to take the child
to visit his various relatives scattered about the country.
Ann Henderson had become the wife of Joseph Burn of
the Red House farm, Wolsingham. She had done better
had she been content with the poor young brakesman;
but she was for a time the most important personage in the
family. She had a strong feeling of kindliness for George,
and when her sister Fanny was no more, she was constant in her hospitality to her nephew. A visit to Wolsingham was the child's highest ideal of happiness; and
when he was there he used to repay his relations for
their goodness by mimicking the peculiarities of his
Killingworth acquaintance. Aunt Burn was in the habit
of giving the little fellow, for his breakfast, fresh eggs with
butter in them. This luxurious fare, so unlike what he
was accustomed to in his father's cottage, appeared to
him in the light of a strange and important discovery,

and it is still remembered how he gravely informed his Aunt Burn that 'when he went home, he'd teach his Aunt Eleanor to eat eggs and butter.'

Another excursion made by the child was to Ryle, where his aunt Hannah Henderson had married Mr. Elliot, a small innkeeper. The time of the year was summer, and as the journey was made on foot, little Bobby and his aunt rested several times on the dusty road, and refreshed themselves at wayside houses of entertainment. A gill of mild 'yell' was the modest order, invariably made by the aunt, and the half pint of drink was always divided between herself and her charge. On reaching Ryle the child found his tongue and impudence, and astounded his relatives by asserting that his staid aunt could not pass an ale-house without entering it. 'Ah! he was a hempy lad,' is the conclusion given amongst his humble relations to nearly all the stories of Robert Stephenson's early life.

Midway in the straggling street of Long Benton, on the right hand of the traveller going from Newcastle to Killingworth township, stands a stone cottage, composed of two rooms—one on the ground-floor, the other upstairs. For many years this has been the village school. At the present time the schoolmaster, in addition to his vocation of teacher, holds the office of postmaster—a fact set forth in bold characters on the exterior of the dwelling. On one side of the school-room, at a rude desk, sit eight or ten boys, whilst on the opposite side are ranged the same number of girls. At one end of the stone floor, between the two companies, sits the instructor, whose terms for instruction vary from threepence to sixpence per week for each pupil. When Robert Stephenson was a little boy, the master of this school was Thomas Rutter. Fifty years ago the village schoolmaster had in many districts a more lucrative business than he enjoys in the present generation. A majority of the surrounding men of business were dependent on a neighbour endowed with 'learning' for the management of their accounts. By keeping the books of prosperous mechanics and petty traders, and by instructing adults bent on self-education, the village schoolmaster found the chief part of his work and payment, apart from his classes for the young. Tommy Rutter, as he is still familiarly called by the aged inhabitants of Long Benton, was both successful and well esteemed.

To Tommy Rutter's school Robert Stephenson was sent, and there he learnt his letters, at the same desk and under the same master as another distinguished child of Long Benton—Dr. Addison, the eminent physician, whose death under mournful circumstances recently created wide and painful sensation. In Rutter's time the girls were taught by Mrs. Rutter in the room upstairs, the ground-floor apartment being filled with lads—the sons of workmen at the surrounding collieries, and of small dealers living in adjacent townships. Many of them had never worn shoe or boot; but, though bare-footed, they were canny, hardy youngsters, and several of them have raised themselves to conditions of prosperity.

The exact year of Robert's entry into Rutter's school cannot be ascertained, but he was quite a little fellow when he first felt his master's cane. The walk over the glebe farm and past the churchyard from the West Moor to Long Benton Street — a distance of about a mile, or a mile and a half — was a long way for him, and Aunt

Nelly used to pity her bairn for having to trudge so far, to and fro. He had not been long at school when the season of harvest came, and Aunt Nelly went out gleaning.

Little Robert Stephenson petitioned his father for leave to accompany Aunt Eleanor and the gleaners. George by no means approved the request, as he argued that he did not pay fourpence, or possibly sixpence, a week for his son's schooling, in the expectation that the young scholar should leave his books at the first temptation.

But the petition was granted in the following terms : —

' Weel, gan; but thou maun be oot a' day. Nae skulking, and nae shirking. And thou maun gan through fra the first t' th' end o' gleaning.'

On this understanding Robert and Aunt Eleanor started for their vagrant toil, but long before sunset the boy was very tired. He kept up manfully, however, and as he trotted homewards at nightfall by the side of his aunt, he, like her, carried a full bag. At the gate of the West Moor cabin stood George Stephenson, ready to welcome them. Quickly discerning the effort Robert was making to appear gallant and fresh, the father enquired :

' Weel, Bobby, hoo did the' come on ? '

' Vara weel, father,' answered Bobby stoutly.

The next day, bent on not giving in, the boy rose early, and for a second time accompanied the gleaners. The poor child slept for hours under the hedgerows ; and when evening came he trotted home, bag in hand, but holding on to Aunt Nelly's petticoats. Again at the garden wicket George received them, with amused look, and the same enquiry :

' Weel, Bobby, hoo did the' come on ? '

' *Middlin*, father,' answered Bobby sulkily ; and, dropping his bag, he hastened into the cottage, and was asleep
in a couple of minutes.

The third day came, and little Robert did his bravest
amongst the gleaners : but the day was too much for
him ; his pride gave in, and on lagging home at nightfall, when he was once more asked by his father, ' Weel,
Bobby, hoo did the' come on ? ' he burst into tears, and
cried, ' Oh, father, warse and warse, warse and warse :
let me gan to school agyen.'

It was not the time then to point the moral of those
last three days, but the next day (Sunday, when even
gleaners rest) the young father took his child under his
arm, and placing him on the knee where he had so often
sat, told him to be a good boy over his book, to leave
hard work of the body for a few years to his elders, and
to thank God that he (unlike his father) was not in childhood required to toil hard all day for a few pence. It
was a sermon fit for a day of rest, and from no lips could
it have come more appropriately than from the lips of
George Stephenson.

Aunt Eleanor sat by, and heard George's paternal
admonition, and was well pleased with its grave and
serious tone. To tell the truth, the Sundays at the West
Moor cottage were not altogether in accordance with
Aunt Eleanor's views. George resolutely declined to
accompany his sister to the meetings of the Wesleyan
Methodists ; and, what to her seemed even worse, he
was by no means a regular attendant at Long Benton
church. Sunday was the day when, walking up and down
the colliery railway, he pondered over the mechanical
problems which were then vexing the brains of all the

intelligent workmen of the neighbouring country. It was his day, too, for receiving friends.

Of George's early associates Robert Hawthorn has been already specially mentioned—and the relations between them have been briefly stated. Whilst George Stephenson and William Locke worked under Hawthorn, they found him an exacting and tyrannical supervisor. They both resented his domination, believing that he was jealous of their mechanical genius, afraid of being supplanted or surpassed by them, and anxious to keep them under. George Stephenson retained for many years a grudge against Hawthorn, but he was too prudent openly to quarrel with the cleverest engine-wright of the district. Slowly advancing himself from the position of a brakesman, whose duty it is simply to regulate the action of a steam-engine, to the higher status of the smith, or wright, who mends and even constructs the machine itself, George stood in frequent need of the counsel and countenance of Hawthorn, then his superior in knowledge, as he was also in age. The practice of the engine-wrights of George Stephenson's Killingworth days was very different from that of the educated engineers of a later date.

John Steele, another of George Stephenson's early and most valued friends, was a man worthy of especial mention ; as his relations with Trevithick, and his ascertained influence on the history of the locomotive, give value to the few particulars that can be picked up with regard to him. The son of a poor North-countryman, who was originally a coachman and afterwards a brakesman on the Pontop Railway, John Steele in his early childhood displayed remarkable ingenuity in the construction of models of machines. His schoolfellows at Colliery Dykes used to marvel at the correctness of 'his imitations of pit-engines,' and remember how in school 'the master could never set him fast' in figures. While he was still a school-lad, his leg was accidentally crushed on the Pontop tramway. After leaving the Newcastle infirmary, where the limb was amputated, he was apprenticed by the proprietary of the Pontop Railway to Mr. John Whinfield, the iron-founder and engineer of the Pipewell-gate, Gateshead. Whilst serving his apprenticeship he attracted the attention, not only of his masters, but also of Trevithick, who in nothing displayed his consummate genius more forcibly than in the sagacity with which he selected his servants and apprentices. In the autumn of 1860, the only sister of John Steele was still living, at a very advanced age, at Ovingham, under the benevolent protection of Mr. T. Y. Hall, of Newcastle, and could re-member that Trevithick invited her brother to leave Whinfield's factory during his apprenticeship and to join him. Steele, however, remained at Gateshead until he had 'served his time,' and then joined Trevithick, during the manufacture of the locomotive constructed by that original mechanician in 1803 and 1804, in the latter of which years the engine won the memorable wager between Mr. Homfray, of Penydarren works, and Mr. Richard Crawshay, of the Cyfarthfa works. Returning from Trevithick's works to Gateshead, Steele, in 1804, built the first locomotive which ever acted on the banks of the Tyne.* This engine was made in Whin-

---

* The facts connected with this engine were brought to light in the columns of the *Gateshead Observer* and the *Mining Journal*. The curious

field's factory for Mr. Blackett of the Wylam colliery ; but owing to the imperfections in its structure, it was never put on the Wylam line, but was used as a fixed engine in a Newcastle iron-foundry. Speaking of this engine, Mr. Nicholas Wood, whose book on Railroads has been copied by all writers on the subject, observes :— 'The engine erected by Mr. Trevithick had one cylinder only, with a fly-wheel to secure a rotatory motion in the crank at the end of each stroke. An engine of this kind was sent to the North for Mr. Blackett of Wylam, but was, for some cause or other, never used upon his railroad, but was applied to blow a cupola at the iron-foundry at Newcastle.' In this statement Mr. Wood fell into a pardonable but not unimportant error. The engine was undoubtedly in all essential points a reproduction of the one already made by Trevithick, with whose name, even more than with those of Leopold, Cugnot, Oliver Evans, or William Murdock, will be associated the practical introduction of the steam locomotive ; but it was made in Gateshead about the year 1804. It is equally certain that John Steele made it, and that when it was finished it ran on a temporary way laid down in Whinfield's yard at Gateshead. John Turnbull, of Eighton Banks, living in 1858, remembered the engine being made, whilst he was serving his apprenticeship at Whinfield's factory. When it was completed, it ran, according to Turnbull's account, backwards and forwards quite well, much to the gratification of ' the quality ' who came ' to see *her* run.'

The subsequent career of John Steele was adventurous. He was employed by the British Government to raise sunken ships ; and, according to his sister's account, received a medal for his efforts to raise the 'Royal George.' Subsequently he went abroad, and having established a foundry and machine factory at La Gare, near Paris, was commissioned to make some engines for several boat companies. His death occurred under painful but characteristic circumstances. Whilst engaged at Lyons in fitting engines on board a boat, he met with Mr. Charles Manby, a gentleman since well known as the Secretary of the Institution of Civil Engineers, but who at that time (1824-5) was engaged in engineering pursuits in France. On the day when Steele's vessel was tried, Mr. Manby took his workmen on board to assist his countryman. On going below, he perceived that the engineman had fastened down the safety-valve, with the avowed intention of 'making her go or bursting her.' Seeing the danger, Mr. Manby and his men hastily quitted the ill-starred vessel. A few minutes later the boiler burst, and by the explosion Steele was killed, together with several important persons of Lyons and many of the spectators on the quays.*

Anthony Wigham, another of George's intimate associates, was the farmer occupying the glebe farm of Long Benton, the cottage-house on which small holding stands within sight of the West Moor cabin. He was a bad farmer, and, as bad farmers usually are, a poor one ; but he had mastered the principal rules of arithmetic, and had a smattering of natural philosophy. George culti-

---

* Minutes of Proceedings Inst. C. E., vol. xii. p. 143.

vated the farmer's acquaintance, and gained from him all
the little knowledge he could impart. The teacher was
in after life amply repaid for his lessons. Bad farming
was in due course followed by commercial failure, and
when the farmer was at a loss where to look for daily
bread, George Stephenson—then grown a rich man—
took him to Tapton House, and, having made him the
superintendent of his stables, treated him kindly to the
last.

Another of George Stephenson's early friends was
Captain Robson, a hale, hearty, manly sailor. His early
life had been passed on board a man-of-war, and he
afterwards became captain of a Newcastle trading
vessel, built for him by his father. Marrying the only
daughter of a prosperous farmer, Captain Robson gave up
sea-life, and became a farmer in Killingworth township.
It was in his house that George discussed his schemes for
the construction of the famous safety-lamp. After again
turning sailor and again relinquishing the sea, the captain
still lives to tell his version of the way in which the
secret of the invention of the lamp was foolishly blabbed
by Dr. Burnet, the colliery-doctor, to his brother-in-law,
Mr. Buddle, the viewer, who, he alleges, speedily conveyed
the information to Sir Humphry Davy. The captain's
story, thoroughly believed as it is by the veteran, is, of
course, not to be relied upon; but it forms an amusing
counterpart to the angry accusations preferred by Sir
Humphry's friends against George Stephenson, of having
surreptitiously possessed himself of the philosopher's
secret.

Hawthorn and Steele, living at a distance, were comparatively rare visitors at Killingworth. George saw

more of them on pay-nights at Newcastle, when he and all
the clever mechanics of the country round met together,
and exchanged views on the difficult 'jobs' then engaging
the attention of the local engine-wrights ; the simple
workmen thus unconsciously creating the earliest and
the finest school of practical engineering.   When, however, either Hawthorn or Steele did make an appearance at the West Moor, the favourite topic was the
possibility of employing steam for purposes of locomotion.   Every word that came from Steele—Trevithick's
pupil and workman, who had himself within six miles
of Killingworth built a machine which, with all its
defects, had actually travelled under the influence of
steam — George Stephenson stored up in his memory.
Steele was never weary of prophesying, that 'the day
would come when the locomotive engine would be fairly
tried, and would then be found to answer.'   No wonder
that George Stephenson caught enthusiasm from such a
teacher.

CHAPTER III.

ROBERT STEPHENSON, THE SCHOOLBOY.

(ÆTAT. 9-15.)

Robert and the Pitman's Picks — 'Mind the Buiks' — George Stephenson's pecuniary Position whilst his Son attended Rutter's School — George appointed Engineer to the Collieries of 'The Grand Allies' — The Locomotive on the Wylam Line — George Stephenson's first Locomotive — His Appointment to the 'Walker Iron-works'—'Bruce's Academy'—The Cost of Robert's Tuition at the School — Robert Stephenson's Reception by his new Schoolfellows — The Boy's delicate Health — The Purchase of his Donkey— John Tate — Rival Safety Lamps—Testimonial and Public Dinner to George Stephenson for his Lamp — Home Gossip—'Throwing the Hammer' — George Stephenson's Views with regard to the Education of his Son — Robert Stephenson's Plan of a Sun-Dial.

AS soon as little Robert was strong enough to help his father, he was put to do such jobs as were suited to his powers. One of his earliest recollections in after life was of having to carry the pitmen's picks to the smith's shop in Long Benton, when they needed repair. This commission he executed on his way to Tommy Rutter's school, and as he returned home he used to bring the implements back. Two years before his death, after his brilliant career of adventure and success, he visited Long Benton with some friends, and pointed out to them the route over the fields, along which he used to trudge laden with the hewers'

implements. But George's chief injunction to his only
child was to 'mind the buiks.' The father was determined that his boy should not commence the real battle
of life, as he had done, unable to cipher, or write, or
even to read.

An erroneous impression exists that George Stephenson denied himself the indulgences appropriate to his
condition in order that he might give his boy a
superior education, and that in sending his son to school
he showed his superiority to most of his fellow-work-
men. He felt personally the disadvantages of a very
defective education, and he determined that his son
should not labour under the same want.

In 1812, on the death of Cree, the engine-wright of the
Killingworth colliery, George Stephenson was appointed
engineer, with a salary of £100 per annum, to the con-
tiguous collieries possessed by Sir Thomas Liddell, Mr.
Stuart Wortley, and the Earl of Strathmore—the 'grand
allies,' as they were called in the neighbourhood. In
addition to this salary, George had the proceeds of
his clock-mending and clock-cleaning business — a much
more important source of gain than has hitherto been
supposed. He not only kept in order the clocks of the
pitmen and superior workmen, but performed the same
service for surrounding farmers. Farmer Robson paid
him half-a-crown for cleaning watch or clock. He was
also regularly employed at a fixed annual sum to
attend to the clocks in the establishments of several
wealthy gentlemen of the vicinity. Moreover, throughout
the term of his Killingworth residence, he lived rent-free
and had his fuel from the pit. During the year, also,
he increased his income considerably by jobs connected

with the repair of machinery. His income therefore amounted in 1812 to about £150. With such means at his command it was only natural that he should give his son the rudiments of education at the village school. Thus in sending Robert Stephenson to Rutter's school, George Stephenson only did as every reputable father of his own station and of similar means in the parish of Long Benton did as a matter of course.

On gaining the important post of engineer to the collieries of the 'grand allies,' George Stephenson's advances towards success became quicker, and at the same time easier. Watchful of all that was going on in the neighbourhood relative to the steam engine, he knew the result of the memorable experiments on the Wylam line, as soon as they were accomplished. On that line it was first proved by Mr. Hedley, the viewer of Mr. Blackett's colliery, that the adhesion* of smooth

---

* 'About this time Mr. Blackett had considerably improved his engines, and by experiments had ascertained the quantity of adhesion of the wheels upon the rails, and had proved that it was sufficient to effect the locomotion of the engine upon railroads approaching nearly to a level, or with a moderate inclination. His railroad was a plate-rail, and would consequently present more friction, or resistance, to the wheels than the edge-rail, and on that account the amount of adhesion would be greater than upon the other rail. Still the credit is due to Mr. Blackett for proving that locomotion could be applied by that means only.' — Mr. Nicholas Wood's *Treatise on Railroads*, third ed. p. 285.

'It was, however, a question of the utmost importance to ascertain if the adhesion of the wheels of the engine upon the rails were sufficient to produce a progressive motion in the engine, when loaded with a train of carriages, without the aid of any other contrivance; and it was by the introduction and continued use of them upon the Wylam railroad that this question was decided : and it was proved that upon railroads nearly level, or with very moderate inclination, the adhesion of the wheels alone was sufficient, in all the different kinds of weather, when the surface of the rails was not covered with snow.

'Mr. Hedley informs us that they first tried by manual labour how wheels on smooth rails would afford sufficient resistance to enable an engine to drag a train of loaded carriages. And it was on that same line, between Wylam and Lemington, that engines with smooth wheels, running on smooth rails, first took the place of horses and oxen for purposes of traffic.

The alacrity with which George Stephenson, the self-taught engineer, comprehended the importance of the Wylam discoveries, and put them in practice upon the Killingworth line, in locomotives of his own construction, which were fully equal in efficiency to those on the Wylam way, attracted general attention to his proceedings. It was seen that he was a man who, with favourable opportunities, would become a distinguished engineer. The Wylam way was laid with plate rails, whilst the Killingworth line had edge rails. George Stephenson therefore built 'the first locomotive engine that propelled itself by the adhesion of its wheels on edge rails.' The first trial of the engine took place on July 25, 1814, with marked success. When the training and antecedents of the young workman (then only thirty-three years of age) are taken into consideration, the achievement seems almost incredible.

Amongst the gentlemen of the neighbourhood who watched the progress and hailed the success of George Stephenson's first engine, no one was more enthusiastic than Mr. Losh, the senior partner of the firm of 'Losh,

---

much weight the wheels of a common carriage would overcome without slipping round upon the rail; and having found the proportion it bore to the weight, they thence ascertained that the weight of the engine would produce sufficient adhesion to drag after it, upon their railroad, a requisite number of carriages.'—Wood's *Treatise on Railroads*, third ed. p. 287.

Wilson, and Bell.'   This highly cultivated gentleman, the fellow-student and friend of Humboldt, survived in a venerable old age in the autumn of 1860, to tell the story of his intercourse with George Stephenson.   With a large capital embarked in the Walker iron-works, as well as in his chemical factories, he saw in the engine-wright a man well fitted to carry out his enterprises and to suggest new ones.   He made overtures to him; and, in the beginning of the year 1815, an arrangement was made that George Stephenson should come to the Walker iron-works for two days in each week, receiving for his services a salary of £100 per annum, besides participation in all profits arising from his inventions.   To secure his good fortune in this compact from all drawback, the 'grand allies,' with proper liberality to an engineer who had served them well, gave him permission to accept Mr. Losh's offer, and at the same time retain his post at Killingworth with an undiminished salary.

George Stephenson, with these two concurrent appointments yielding him a clear £200 per annum, besides perquisites and the participation in profits reserved to him by Losh, Wilson, and Bell, began to feel himself a rising man.   Industrious as ever, he retained his clock-cleaning business; and he had made some not unimportant savings.   A prosperous mechanic, with a good income, unmarried, and with brighter prospects opening before him, could not think of giving his only child no better education than that which a village schoolmaster imparted to the children of ordinary workmen.

It was no part of his plan to bring up his son with an expense and refinement unusual in his station, but he wished to educate him in accordance with the rules of his rank.   He placed him, therefore, when he was nearly twelve years old, as a day-pupil in an academy at Newcastle, kept by Mr. Bruce.

The friend and biographer of Dr. Hutton, and the author of several educational works of great merit, Mr. John Bruce had raised his school to such excellence that it then ranked higher than the Newcastle grammar-school, where Lord Stowell, Lord Eldon, and Lord Collingwood received their early instruction.   The ' Percy Street Academy '— as Mr. Bruce's seminary was and still is called—was then attended by more than a hundred pupils, who might be described as a good style of ' middle-class boys.'   Some few were the sons of the minor gentry of the vicinity, but the majority were the sons of professional men and traders of Newcastle and Gateshead.   Not one half of the boys learned either Greek or Latin.   Amongst those who did not receive classical instruction was Robert Stephenson, who entered the school on August 14, 1815, and remained there four years.   During that time, the whole sum paid for his education fell short of £40.   The expenditure, therefore, for a father in George Stephenson's circumstances, was sufficient and appropriate, but nothing more.

On Robert Stephenson's appearance at the Percy Street academy he had to encounter the criticisms of lads who regarded him as beneath them in social condition. ' A thin-framed, thin-faced, delicate boy, with his face covered with freckles,' * dressed in corduroy trowsers and a blue coat-jacket, the handiwork of the tailor

---

* Such is the description of him given by a Newcastle gentleman who distinctly remembers his first coming to Bruce's school.

employed by the Killingworth pitmen, the new-comer presented many marks for play-ground satire. On his shoulder he carried a bag containing his books and a dinner of rye-bread and cheese. The clattering made by the heavy iron-cased soles of his boots on the school floor did not escape the notice of the lads. Mr. Bruce was on the look-out to see that he was not improperly annoyed; but there was no occasion for the master's interference. In Robert's dark eyes there was a soft light of courtesy that conciliated the elder boys. When they entered into conversation with him, however, they could not refrain from laughing outright. Gruff as their own voices were with Northumbrian 'burr,' they were unused to the deep, guttural pit-intonations with which Robert expressed himself. It was no slight trial to a sensitive child just twelve years old to find himself the object of ridicule. Puzzled as to what he had said that was ludicrous, and deeply mortified, he turned away, and kept silence till the business of school-hours commenced.

At first Robert Stephenson walked to and from school —a distance in all of about ten miles; and this labour disinclined him for joining in the sports of the play-ground. At dinner he held no intercourse with his schoolfellows; for while they consumed the more luxurious fare provided for them by Mrs. Bruce, he ate the inexpensive provision put into his satchel by Aunt Eleanor, or partook of the frugal fare of an uncle's family. Gradually, however, he became a favourite with the lads. But it soon became clear that Robert Stephenson was not strong enough to bear the long walk each night and morning. He was liable to catch cold, and the tendency it had to strike at his lungs made his father apprehensive

that tubercular consumption might attack him. At this
time, too, the boy was afflicted with profuse nightly
perspirations, to obviate which the doctors made him
sleep on a hay mattress. A step more likely to do good
was taken by George Stephenson, who purchased for
the boy a donkey, which was for years the pride of
Long Benton. Robert had for a long time been in possession of a dog and a blackbird, which he used to aver
were the cleverest inhabitants of the village. His new
acquisition gave him lively satisfaction, and he was prouder
of it than he was in after life of any horse in his stable.
To spare his ' cuddy,' he used, in fine weather, to walk
and ride to school on alternate days.

John Tate (in 1860 the foreman blacksmith at the
colliery,) the son of George Stephenson's old friend,
Robert Tate, formerly the landlord of the Killingworth ' Three Tuns,' was in early boyhood the familiar companion of Robert Stephenson. The two lads
had many a prank together. Shortly before Robert
left Rutter's school, they were out birds'-nesting, when
Robert fell from a high branch of a tree to the ground,
and lay for a minute stunned. On recovering his con-
sciousness, he experienced so much pain on moving one
of his arms that he nearly fainted. ' My arm is broken,
John Tate,' the little fellow said quietly ; ' you must
carry me home.' Luckily John Tate had not far to
carry him. In due course the broken arm was set ; but
throughout the operation, and indeed from the time
when he told John Tate to carry him home until he was
asleep, he did not utter a cry of pain. A child of eleven
years who could evince such fortitude was clearly made
of the right stuff.

The first half year of Robert Stephenson's career at the Percy Street academy was an eventful one with his father. · It saw the invention of the Geordie safety-lamp, and the outbreak of that contest between Sir Humphry Davy and the Northumbrian engine-wright, in which the latter unquestionably displayed the greater dignity and moderation. George Stephenson's first lamp was tried on October 21, 1815. In the Northumbrian coal fields three lamps are used more than any of the others which inventors have contrived for the protection of the miner,— Dr. Clanny's lamp of the year 1813, and the lamps invented two years later by the scientific reasoner Sir Humphry Davy, and the practical mechanician George Stephenson. The principle in each of these last-named lamps is identical, but the two originators arrived at it by very different processes. To decide on the respective merits of these lamps is no part of this work. Each has its supporters ; and the partizans of a particular kind of ' safety-lamp ' are scarcely less vehement and uncharitable in their zeal, than are the defenders of a particular school of religious opinion. In the mines where ' the Clanny' is used, nothing but ' the Clanny' has a chance of trial, or a good word. The same is the case with ' the Davy ' and ' the Geordie.' One thing, however, is certain. An efficient and luminous safety-lamp is still to be invented. It is amusing to hear the virtuous indignation of those who, never having visited the narrow passages of a coal mine, vehemently condemn the fool-hardiness and perversity of miners who prefer the candle to the lamp. So dim a ray is emitted by ' the Davy ' or ' the Geordie,' it is far from wonderful that underground toilers should regard them as

obstacles to industry rather than as agents for the preservation of life.

With regard to George Stephenson and his invention,
the time has come for the final sweeping away of a
fiction. The true nobility of the elder Stephenson is
only insulted by those who would surround it with the
vulgar glare of melodramatic heroism. Amongst the
many anecdotes by which indiscreet eulogists have hoped
to exalt the fame of a remarkable man, is the story that
George Stephenson, to test the worth of his lamp, took
it on the memorable night of October 21, 1815, into
the foulest part of a foul mine, at the peril of instant
destruction. Had such a risk been necessary to preserve the lives of his fellow-creatures, such conduct
would have entitled him to endless praise for self-
sacrificing intrepidity. But as he knew there was no
need to incur such danger, the act attributed to him
would have deserved no commendation. Wilfully and
deliberately to encounter extreme peril, with the full
knowledge that it is needless, is the part of a fool—not of
a hero. Whatever may be George Stephenson's claim to
be regarded as the latter, he certainly had nothing in
common with the former. The important experiment,
which has been so greatly misrepresented, was made
on a certain insulated quantity of gas, and under circumstances that precluded the possibility of serious
disaster. Mr. Nicholas Wood, the well-known writer
on Railroads, at that time the 'viewer' of the colliery,
assisted at that trial, and says, 'the box, or cabin, in
which the lamp was tried was not of such dimensions as
would, if an explosion had taken place, have produced
the effect described; as only a small quantity of gas was

required, and we had had sufficient experience not to employ more gas than was necessary : at most, an explosion might have burnt the hands of the operator, but would not extend a few feet from the blower.'

To George Stephenson one of the best consequences of his invention was the quarrel which it provoked between his friends and the supporters of Sir Humphry Davy. The coal-owners of the district formed themselves into two parties. A newspaper war was waged, in which the advocates of Stephenson were altogether victorious. The partizans of Sir Humphry gave him as a reward for his invention £2,000, awarding to George Stephenson 100 guineas for the lamp they professed to regard as a clumsy contrivance, if not an imitation. This award was officially communicated to George Stephenson by his dogged, but honest, opponent, Mr. Buddle.

To make head against this demonstration of Sir Humphry's friends, George's supporters got up another testimonial, amounting to £1,000. A part of this sum was expended on a silver tankard* which, together with the balance of the money, was presented to the inventor of 'the Geordie,' after a public dinner given at the Assembly Rooms at Newcastle. The chair was taken by George's

---

* The tankard was inscribed — 'This piece of plate, purchased with a part of the sum of £1,000, a subscription raised for the remuneration of Mr. George Stephenson for having discovered the fact, that inflamed fire-damp will not pass through tubes and apertures of small dimensions, and having been *the first* to apply that principle in the construction of a safety-lamp, calculated for the preservation of human life in situations formerly of the greatest danger, was presented to him at a General Meeting of the Subscribers, Charles John Brandling, Esq., in the chair, January 12, 1818.' Among the numerous pieces of 'presentation plate' on Robert Stephenson's sideboard in after days, THE TANKARD was always the most prized.

hearty patron, Mr. Brandling, of Gosforth Hall; and of course George, as the distinguished guest of the night, had to return thanks for the honour done him. In his palmiest days George Stephenson was not an orator, although when he spoke on subjects which he thoroughly comprehended he expressed himself in a plain, sensible, and terse manner, which carried conviction of his sincerity and of the truthfulness of his narration.

Sorely did he stand in need of eloquence when he stood up in the Newcastle Assembly Rooms, and addressed a company of wealthy merchants and enlightened gentlemen. His speech he had learnt by heart, having composed it and written it out with great care. Fortunately, this interesting document, which ought to be committed to the custody of the Newcastle Literary and Philosophical Institution, has been preserved, and a fac-simile is given in the Appendix. The speech ran thus : —

Sir,—In Receiving this valuable present which you and the Gentleman of this Meeting has bean pleas'd to present me with this day I except with Gratitude But permit me to say valuable as this present is and gratefull as I feal for it I still feal more by being honour'd by such and highley respectable meeting the Gentlemen of which having not only rewarded me beyond any hopes of mine for my endeavours in constructing a safity Lamp but has supported me in my claims as to priority in my invention to that of that distinguished Pholosipher S H Davy. For when I conseder the manner that I have been brought up and liv'd the manner of which is known to many of the Gentleman present and when I consider the high station of S H. Davy his high Charactor that he holds among society and his influence on scientific men and scientific bodys. all of which Sir lays me under a Debt of Gratitude to the Gentlemen of this meeting which Gratitude shall remain with me so long as ever I shall live. I shall conclude, sir, with my heart felt thanks to the Gentlemen of this meeting for their great reward thare support in my struggle with my competitor and hear I beg leave to thank in particular R Brandling, Esqr. which I trust the Gentleman of this meeting will give me Credit for. for I beleive this meeting knows well the active part he has taken in my behalf And I hear do thank him publicly for it.*

Keeping close to the letter of this programme, he acquitted himself creditably, but at a family gathering where the great event of the dinner was discussed in all its bearings, he confessed that his embarrassment whilst he delivered the oration was so great, that his face seemed to him 'all on fire.' 'Oh, Grace,' he said to his sister-in-law Grace Henderson, who had become the wife of Bartholomew Twizell, 'if thou could but ha' seen ma meeting so many gentlemen at the 'Sembly Rooms, thou maught ha' lit a canle at ma face.' On this, Jane, another married sister-in-law, laughed, and made a joke at his rise in life. 'Noo thou 'll be for having a bra' ruffle to th' shirt, and then thou 'll be looking doon on a' th' own frien's.'

'No, Jane,' he answered slowly and seriously, ' thou 'll nivar see no change in ma.'

---

* It has been thought right to retain the faults of orthography and grammar to be found in this and other of George Stephenson's writings incorporated in this work. It is desirable that everything relating to such a man should be known, his weakness as well as his strength. It is a fact to be pondered over, that with his powerful intellect and resolute will, George Stephenson to the last could neither write grammatically nor spell correctly, but had to rely on his secretary. Whilst he was braking the ballast engine at Willington Quay, he borrowed a grammar of Mr. John Dobson, still a distinguished architect at Newcastle. He could not, however, master its secrets, and in a few days brought the book back, saying, 'I oonderstond tha vow'ls, but I canna gat hold o' tha verbs.'

At the narration of this story nearly three years since, more than one of George's humble kin who were present bore testimony that 'he never did change—he was always the same—riches made no difference in him towards his poor relations.'

Whilst George Stephenson steadily progressed in his professional career, his son continued his attendance at Bruce's school. He did not figure conspicuously in the Percy Street play-ground, but at home he displayed no less physical than mental energy. Every evening his father kept him hard at work over the tasks set him at school, and over plans of steam-engines and other mechanical contrivances. The neighbours sometimes thought George was an 'o'er strict father,' and pitied the poor boy who was kept so close to his books. Robert, however, had leisure for amusement. Every autumn he and his friends stripped of fruit the best trees in Captain Robson's orchard. Like his father, too, the boy excelled in athletic sports, throwing the hammer and putting the stone with skill and force.

In throwing the hammer — a favourite sport with Northumbrian workmen—the thrower stands with his legs wide apart, when, putting his arms behind his back, and grasping the hammer by the handle with both hands, he casts it forwards between his legs. Apart from the muscular force employed, the knack greatly consists in letting the hammer go at the right moment. Relinquished too soon, the missile strikes the ground close at the player's feet; retained after the proper moment, it is apt to rise up into the thrower's face. In his sixteenth year, Robert was engaged at this pastime, and made the mistake of keeping the hammer too long in hand. The consequence was that the ponderous implement, weighing a little under 28 lb., rose, struck him on the forehead, and laid him flat and perfectly stunned upon the ground. John Tate witnessed the accident; but on the following day he saw Robert throwing the hammer with as much resolution as ever.

Robert's schoolfellows at the Percy Street academy failed to detect in him any remarkable signs of talent, and some of them still express their astonishment at his subsequent scientific acquirements and professional achievements.

Before leaving Robert Stephenson's school-life, we may remark, that his father's experiences and difficulties were the measure of what he thought requisite for the instruction of his son. The subtler influence of letters and the more valuable results of culture were matters about which George Stephenson thought little. Learning he regarded in a strictly utilitarian sense, as an engine necessary for the achievement of certain ends. His ambition was to be a skilful engineer, and a perfect man of business; and in his efforts to achieve this ambition he found two perplexing obstacles in his ignorance of mathematics and his inability to write with facility, or logical exactness. What he desired to be himself, that he also wished his son to be. Robert Stephenson should be an engineer and a director of labour; but he should not have his bravest exertions baffled by defective knowledge. In this spirit George caused his son to learn French, because it would be useful to him in business.

Up to the time when he left Bruce's school, Robert did not exhibit any marked enthusiasm for the pursuits in which his father was most warmly interested. Possibly George Stephenson was too urgent that he should prosecute the study of mechanics, and by continually goading him to work harder and harder ' at his buiks' gave him a transient distaste for subjects to which he was naturally inclined. As a member of the Philosophical and Literary Society of Newcastle, Robert brought home standard popular works and encyclopædic volumes treating of natural science and of inventions. These books his father read and compelled him to read ; but the labour went very much against the boy's grain.

The earliest ' drawing ' by Robert Stephenson's hand of which there is any record, was that of a sun-dial, copied from Ferguson's ' Astronomy,' and presented by the lad to Mr. Losh, in the year 1816, in token of his gratitude to him as his father's benefactor. This drawing set the father and son on another work—the construction of a real sun-dial, which, on its completion, was fixed over George's cottage door, where it still remains, bearing the date, ' August 11th, MDCCCXVI.'

A good story is told of ' the hempy boy,' who dearly loved mischief. From the meadow before the West Moor cabin he sent up his enormous kite, reined in by copper wire instead of string, the copper wire being insulated by a piece of silk cord. Anthony Wigham's cow, peacefully grazing in the meadow, was first favoured with a smart dose of electricity, one end of the copper wire being brought down to the top of the animal's tail. Standing at his cottage window, George Stephenson watched the discomfiture of his neighbour's cow in high glee ; but when the operator, ignorant whose eyes were upon him, relinquished the torture of the ' coo,' and proceeded to give his father's pony a fillip with the subtle fluid, George rushed out from his cottage with upraised whip, exclaiming, 'Ah! thou mischeevous scoondrel — aal paa thee.' It is needless to say that Robert Stephenson did not wait to ' be paid.'

CHAPTER IV.

ROBERT STEPHENSON, THE APPRENTICE.

(ÆTAT. 15-20.)

Robert Stephenson leaves School—He is apprenticed to Mr. Nicholas Wood—George Stephenson lays down the Hetton Colliery Railway— Father and Son—Robert's Economy in his personal Expenses —The 'Three Tuns'—The Circumferentor—George Stephenson's increasing Prosperity — His Second Marriage — He builds the 'Friar's Goose Pumping Engine' — He embarks in a small Colliery Speculation — The Locomotive Boiler Tubes of the Messrs. James — Explosion in the Killingworth Mine — George Stephenson's First Visit to Mr. Edward Pease — Robert Stephenson and his Father survey the Stockton and Darlington Line — Robert Stephenson's First Visit to London — His delicate State of Health — Survey for the Second Stockton and Darlington Act — Robert Stephenson goes to Edinburgh — Professor Leslie's Testimonial — Letters written at Edinburgh by Robert Stephenson to Mr. Longridge — Robert Stephenson accompanies Professor Jamieson on a Geological Excursion—George Stephenson's Letter to his friend Locke — Robert Stephenson and his Father visit Ireland — Robert Stephenson's Letters from that Country.

LEAVING school in the year 1819—the year in which his father commenced the construction of his first line of railway, the Hetton Colliery line—Robert Stephenson entered on his duties as apprentice to Mr. Nicholas Wood, the mining engineer, who was at that time the viewer of the Killingworth and other adjacent collieries. During his apprenticeship, he had therefore to concern himself with the internal working of the

mines to which his father was engine-wright. The
father and son now came closer together, and strengthened the firm league of confidence and affection which
bound them throughout life. There was between them
far less difference of age than usually exists between
father and son, George Stephenson being only twenty-two
years his boy's senior. When Robert Stephenson was a
young man, his father was still only at the entrance of
middle life ; indeed, the latter was, in some respects, a
young man even to the last, anxious for fresh knowledge, capable after a struggle of relinquishing old errors,
and moreover endowed with high animal spirits.

Robert Stephenson was apprenticed to Mr. Nicholas
Wood for three years, and during his apprenticeship he
manifested that quiet resolution and genuine modesty
which characterised him even when he became the
leader of his profession. He worked very hard, and
lived with careful economy. George Stephenson saw
clearly that the only chance he had of reaping a rich
harvest from his own and his son's intellects, lay in saving
and putting by out of his yearly earnings, until he
should be in a pecuniary position to embark in business
as a manufacturer as well as an operative engineer. He
knew well that the inventor without capital makes others
rich, whilst he himself starves and is neglected. His
great object, therefore, was to accumulate funds in order
that he might enter into business as a manufacturing
engineer.

At this period of his life Robert never spent a penny on
any article whatever, until he had put to himself Sydney
Smith's three questions—Is it worth the money? Do
I want it? Can I do without it? Once every fortnight

Mr. Wood, as head viewer, used to descend the Killing-worth mine in discharge of his regular duties. The hour at which he 'left bank' was nine o'clock, punctual to the minute, and Robert always accompanied his master. At mid-day, when the morning's work was over, Robert and the under viewer, hot and fatigued, used to enter the 'Three Tuns'—a small, thatched, three-roomed beer-house, long since pulled down—and take refreshment. When herrings were in season, the ordinary repast of each was 'a herring, a penny roll, and a glass of small beer.' Young gentlemen, serving their pupilage under distinguished engineers, would sometimes do well to think of Robert Stephenson's two-penny-halfpenny meals.

About two years before Robert Stephenson's death, a workman of Washington village found in a collection of old stores a circumferentor, or mining compass. It was unusually large — even for a circumferentor made forty years since. The brass stand and measuring-plate had long been obscured by corrosion ; and it was not till the latter had been well scoured and polished that it revealed the inscription, 'Robert Stephenson fecit.' The workman, on reading these words, brought the instrument to the works of Robert Stephenson and Co., Newcastle, and left it with Robert Stephenson's friend and partner — the late Mr. Weallens. At his next visit to Newcastle, Mr. Stephenson's attention was directed to the instrument, when at the sight of his long-forgotten work, he exclaimed with emotion, 'Ah, that circumferentor was measured off at Watson's Works, in the High Bridge.* I made it when I was

---

* i. e. the High Bridge of Newcastle.

quite a lad—when I was Wood's apprentice—when I had
but little money, and could not afford to buy one.'

Whilst Robert Stephenson was serving his apprenticeship, events were being crowded into his father's life. In
1819, George Stephenson began to lay down the Hetton
Colliery Railway, which was finished in 1822. He could
now afford to indulge in romance. Elizabeth Hindmarsh, his first love, was still unmarried. When her
father drove the young brakesman from his door, she had
vowed never to have another husband, and that vow she
kept. The time was now come for her constancy to be
rewarded. The poor brakesman had made himself ' a
man of mark,' and — a more important matter still in the
estimation of some of his canny north-country friends —
had made himself a ' man of substance.' ' The grand
allies,' in their conduct towards their agent, showed a
liberality becoming their rank, wealth, and name. In the
same way that, years before, they had given him two out
of every six working days, allowing him to devote them
to the service of Messrs. Losh, Wilson, and Bell, so they
now also permitted him to act as engineer to the Hetton
Coal Company, for the construction of the Hetton Railway, without making any diminution in his salary. Thus
during the three years in which he was laying down the
Hetton line, George Stephenson had three concurrent
appointments. His savings were by this time considerable, and were invested at good interest and on good
security. Mortgage on land at five per cent. interest
was at that time George's notion of a sufficiently profitable and safe investment, and on such terms he had
for some years lent £1,300 to a gentleman in the neighbourhood of Darlington. So George Stephenson (no

longer a poor brakesman) again paid his addresses to
the woman whose love he had won twenty years before;
and he married her in the same church where he had
wedded his 'old maid' bride, Fanny Henderson. The
ceremony took place in the parish church of Newburn on
March 29, 1820, the bridegroom's son, Robert, being one
of the attesting witnesses.*

As soon as the wedding festivities were at an end,
George Stephenson went back to his work and his cottage
at Killingworth. Still pursuing his prudent course, he
made no difference in his plan of life; nor, to her
lasting honour be it said, did Mrs. Stephenson wish
him in any respect to alter it. Never did woman more
cordially devote herself to the interests of her husband and husband's child. Entering the Killingworth
cottage, which 'Aunt Eleanor' had left to marry an
honest and well-reputed workman, she gave a beauty
and completeness to her husband's life which it had
previously wanted. Of this excellent lady mention will
be made in subsequent pages. Possibly his step-mother's
tastes turned Robert Stephenson's attention to music.
He purchased a flute, and acquired so much profi-
ciency on the instrument, that he was permitted to act
as flutist in a band, which, instead of an organ, took
part in the religious services of Long Benton Church.

---

* Copy of the record of George
Stephenson's second marriage, in
the Newburn Register : —

'George Stephenson, of the parish
of Long Benton, widower, and Elizabeth Hindmarsh, of this parish, spin-
ster, were married in this church by
license, with consent of ———, this
twenty-ninth day of March, in the
year One thousand eight hundred
and twenty,

'By me, J. Edmonson, vicar.
'In presence of—
 'Thomas Hindmarsh.
 'Robert Stephenson.
  'George Stephenson.
  'Elizabeth Hindmarsh.'

At the same time that George Stephenson was laying
down the last rails of the Hetton Colliery Railway, he was
busy in constructing for Messrs. Losh, Wilson, and Bell, a
pumping engine, of hitherto unusual dimensions, known
as the Friar's Goose Pumping Engine,* which aided in
'the winning' of the famous Woodside coals. The
opening of this mine commenced in 1820, and the first
cargo of coals was shipped November 21st, 1824. The
cost of winning was about £22,354; and George
Stephenson's engine, which speedily became famous
throughout the Northumbrian coal district, commenced
pumping in July 1823. The increase of reputation
which the engineer gained by this achievement was of
great service to him. He had also another important
undertaking on his hands. In conjunction with Thomas
Mason, he took a lease of the Willow Bridge colliery
for twenty-one years, the two partners embarking in
the undertaking £700 in equal shares. The deed of
partnership was signed December 5th, 1820.

Another incident of importance marks this period of
George Stephenson's career. Anxious to improve the
locomotive engines, for which he and Mr. Losh had taken
out letters patent, George and his copatentee resolved to
introduce into their boilers the tubes recommended by

---

* The following particulars concerning the 'Friar's Goose Pumping
Engine,' furnished by Mr. Losh, are
valuable : —
'Friar's Goose Pumping Engine.
Commenced pumping in July, 1823.
Diameter of cylinder 72⅝ inches;
length of stroke, ditto, 9 feet; length
of pit, ditto, 7 feet 2 inches. Two sets
of pumps attached to the out end of
the main beam, and one to inside, by
diagonal spear to quadrant in pit,
about 7 fathoms down from surface.
Three sets of pumps in bottom, each
set 16½ inches diameter, and length
of sets about 50 fathoms. Average
quantity of water per minute, 1,000
gallons.
'Tyne Main Colliery, Aug. 29,
1860.'

Messrs. William James and William Henry James, giving those gentlemen a share in their patent rights in return for the permission granted them ' to adopt any improvements, and the introduction of tubes to their boilers, as contained in the letters patent of William Henry James, son of the said William James, as granted to him in the reign of his present Majesty.' The agreement between William Losh and George Stephenson on the one part, and the Messrs. James on the other, bears date September 1, 1821. These tubes must not, however, be confounded with the multitubular boiler, which ultimately decided the triumph of the locomotive. Almost countless unsuccessful experiments were made, before Mr. Henry Booth (with the concurrence of the Stephensons) produced his beautiful arrangement. The agreement of September 1st, 1821, is of interest, as it gives a date when George Stephenson was intent on increasing the heating surface of his boilers by the introduction of tubes, and also preserves the reputation of two other inventors, whose services to the locomotive ought not to be forgotten, although they have been exaggerated by indiscreet friends.

Robert Stephenson's work during his apprenticeship was not only hard but hazardous. On one occasion when he was accompanying his master, Mr. Nicholas Wood, and Mr. Moodie, the under-viewer, through the passages of the Killingworth mine, by the aid of ' the Geordie's ' dim ray, they grew impatient of the darkness, and lighted a candle. The spot was more foul than the viewer supposed, and an explosion instantly ensued. Mr. Wood was picked up from the ground bruised, bleeding, and stunned. Robert Stephenson and Mr.

Moodie escaped unhurt; but the alarm of such an escape strongly impressed the former with the value of his father's invention.

The lad's apprenticeship had not expired, when he made trial of a safer, but not less laborious, occupation. On April 19, 1821, the same day on which the royal assent was given to the first Stockton and Darlington Railway Act, George Stephenson went over to Darlington, accompanied by Mr. Nicholas Wood, for the purpose of soliciting Mr. Edward Pease, the chief projector of the new line, to secure for him the job of making the railroad.

In consequence of this interview with Mr. Pease, George Stephenson was employed by the Stockton and Darlington Company to make a careful survey of the route, for which the Act had been obtained. This survey was made in the autumn of 1821, and certain modifications and changes of the line were proposed by the engineer. To carry out these proposals, a new Act (the second Stockton and Darlington Railway Bill) was, after renewed opposition, obtained in 1823; and George Stephenson was forthwith instructed to form the line in accordance with the new Act, receiving for his salary as the Company's engineer-in-chief £300 per annum. In making the survey of 1821, Robert Stephenson, then just eighteen years of age, accompanied and assisted his father.

Before entering on the survey, Robert Stephenson made a trip to London. Easy and secure in his circumstances, his father gave him a purse of money and a holiday. It was the first time in his life that he had been more than a day's journey from Killingworth, and the prospect of visiting the capital greatly excited him. Having reached London, the tall slight boy, still dressed in ill-fitting coarse garments made by the pitmen's tailor, hastened from place to place. The journal still exists in which he began to take notes of all he saw. Before he had been in town many days the diary was discontinued; but enough was written to show that he was still unable to spell correctly. He went to St. Paul's, the Custom House, the London Water Works, 'Sommersite' House, and to an exhibition of a model of an Egyptian tomb sent home by Belzoni.

The visit to London was a short one; and when it was over, Robert Stephenson returned to Killingworth to resume his work in the coal-mines. But by this time he had found the labour of a viewer exhausting as well as perilous. His lungs were weak and manifested symptoms of tubercular disease. He welcomed, therefore, the change to a more healthful occupation now offered to him, and in the early autumn assisted his father and Mr. John Dixon in making the survey for the second Stockton and Darlington Railway Act. He heartily enjoyed the work. Spending the entire day in the clear balmy air, eating frugal meals of 'bread, butter, milk, and potatoes' under sheltering hedgerows, and lodging by night in roadside inns, George Stephenson and his assistants made holiday of their toil.

Mr. Joseph Pease of Darlington, then a young man, was a frequent attendant on the party, and remembers well the animation with which George and Robert Stephenson conversed at the top of their voices, in a scarcely intelligible Northumbrian brogue, on the difficulties of their undertaking. The 'slight, spare, bronzed boy,' as Mr. Pease recalls the Robert Stephenson of 1821, often supported his arguments with a respectful mention of Mr. Bruce's opinions; and to the authority of the worthy schoolmaster, George Stephenson invariably paid marked, and almost superstitious, homage.

When the survey was completed, and the map was plotted, Robert Stephenson's name was put upon it as 'the engineer,' and no mention was made of his father. This was done at George's particular direction; and a more affecting instance of paternal devotion it would be difficult to imagine.

In consequence of being thus designated engineer, Robert Stephenson had to make a second visit to London, and this time not for the purpose of inspecting the Tower and St. Paul's Cathedral, but that he might be examined by a parliamentary committee on an affair of great commercial importance.

Before making his first public appearance as engineer of the Stockton and Darlington Railway, Robert Stephenson resided for a few months in the university of Edinburgh. Several gentlemen who came in contact with him during the survey for the line had been so struck with his natural force of intellect that they represented to his father the propriety, and indeed the imperative duty, of giving him a college education. George Stephenson could, as far as money went, have well afforded to send him to Cambridge, but it was not his wish to 'make his son a gentleman.' Such were his own words. 'Robert must wark, wark, as I hae warked afore him,' the father used to say. Finding, however, that his son could reside at Edinburgh, and attend the classes for a comparatively small sum, he allowed him to go to that university for one term, a space of time that was, in all, something less than six

months. This permission was accorded in the October
of the year 1822, and forthwith Robert Stephenson
started for the Scotch capital. As the date of his
residence in Edinburgh has been misstated, so also has
the importance of it been exaggerated. To call it
by the imposing title of a ' university education ' would
be to mislead the reader. Brilliant as the assembly
of professors in Edinburgh then was, the educational
system of the university was faulty, and the students
were allowed to pursue their own courses, without disci-
pline, and in some cases without encouragement. Robert
Stephenson certainly worked hard whilst he was at
Edinburgh, but his stay there was too short for efficient
study. He was, however, resolute in his attendance at
lectures, and he even declined to enjoy for an ,hour the
society of Mr. Joseph Pease (who paid him a flying visit)
in order that he might be present at the address of the
Professor of Natural Philosophy.

After the term he accompanied Jamieson on a geological
excursion. The students who were permitted to attend
the Professor on such trips walked with knapsacks on their
backs, and led the same sort of wild vagrant life which
Robert had more than a year before enjoyed during
the railway survey. To the last he retained a lively
recollection of this expedition; and as late as 1857, on
passing in his yacht an imposing headland of the northern
coast, he told his friends that, ' as a student on a tour with
Professor Jamieson, a quarter of a century before, he had
examined the structure of the cliffs.' ' The Professor,' he
added, ' on such occasions mounted a hillock and described the geological formation of the surrounding rocks,
illustrating his lecture by reference to the face of nature

as his black-board, while we lads stood round the good old man with a pleasure which I can never forget.'

It has been erroneously stated that Robert Stephenson bore off at Edinburgh 'most of the prizes of the year.' The fact is, that he did not gain a single university prize, in the sense in which an university man would use the term.

Professor Leslie, however, was in the habit of presenting periodically a book to the student attending his class with whom he was most pleased. According to the character of the pupil to whom it was presented, it was sometimes a tribute to moral worth as well as scientific attainments. In the case of Robert Stephenson, the Professor's testimonial was awarded in recognition of the ability displayed by the pupil in answering certain mathematical questions in the regular weekly examination papers.

The following letters written by Robert Stephenson to his early friend and adviser, Mr. Michael Longridge, during his brief stay at Edinburgh, will give the reader an insight into his life in the university. The first of the three was written soon after his arrival in the capital of Scotland, and whilst he was making a first acquaintance with the Professors.

Edinbro': Nov. 20, 1822.

SIR,—Not having received the books, as you intimated, I begin to be apprehensive of their safety. If you have not sent them off yet, I hope you will not be long. I met with very kind reception from Mr. Bald, who introduced me to Dr. Brewster, Professor Jameson, and some other professional gentlemen. He gave me two tickets, one for the Wernerian Society, and one for the Royal Society, and desired me particularly to call and have any book out of his library that I might want. Mr. Jameson seems to be a very intelligent man, and I think him and I will soon be friends. My father would likely inform you of my intercourse with Dr. Hope. He seemed much interested about the lamps, and desired me to give him every information relative to them.

I remain, Sir, yours sincerely,

R. STEPHENSON.

M. Longridge, Esq.

The tone of the next letter, penned a fortnight after the preceding epistle, is less cheerful.

SIR,—I would have sent my Lectures ere now had they contained anything new. Mr. Jameson's Lectures have hitherto been confined chiefly to Zoology, a part of Natural History which I cannot say I am enraptured with ; nor can I infer from many of his Lectures any ultimate benefit, unless to satisfy the curiosity of man. Natural historians spend a great deal of time in enquiring whether Adam was a black or white man. Now I really cannot see what better we should be, if we could even determine this with satisfaction ; but our limited knowledge will always place this question in the shade of darkness. The Professor puzzles me sadly with his Latin appellations of the various divisions, species, genera, &c., of the animal kingdom. He lectures two days a week on Meteorology and three on Zoology. This makes the course very unconnected.

I have taken notes on Natural Philosophy, but have not written them out, as there has been nothing but the simplest parts, and which I was perfectly acquainted with. Therefore I thought I might spend my time better in reading. I shall send you them when he comes to the most difficult parts. Leslie intends giving a Lecture on Saturdays to those who wish to pursue the most abstruse parts of Natural Philosophy. I have put my name down for one. of those : he gives questions out every Friday to answer on the Saturday. I have been highly delighted with Dr. Hope's Lectures. He is so plain and familiar in all his elucidations. I have received the books all safe.

The next letter, written in the April of 1823, marks the time when the writer's brief stay at the university was brought to a close, and also indicates with exactness the subjects to which he directed his attention during the period.

Edinburgh : April 11, 1823.

SIR,—I wrote home on the 5th, but from yours it appears my father would be set off for London before the arrival of my letter, in which I desired him to send me a bill for £26. I should feel obliged if you will send me it at your first convenience, as I am rather in want of it at present.

The Natural History finishes next Tuesday. The Natural Philosophy on Friday the 18th. Chemistry finishes on the 27th or 28th.

I have been fortunate in winning a prize in the Natural Philosophy class, for some mathematical questions given by Professor Leslie relative to various branches of Natural Philosophy.   I remain, Sir,

Yours very sincerely,
ROB. STEPHENSON.

Mich. Longridge, Esq.

The following letter, written by George Stephenson to his friend William Locke, during his son's brief sojourn at Edinburgh, will be read with interest :—

March 31, 1823.

DEAR SIR, — From the great elapse of time since I seed you, you will hardley know that such a man is in the land of the living. I fully expected to have seen you about two years ago, as I passed throw Barnsley on my way to south Wales but being informed you was not at home I did not call I expect to be in London in the course of a fortnight or three weaks, when I shall do my self the pleasure of calling, either in going or coming. This will be handed to you by Mr. Wilson a friend of mine who is by profeshion an Atorney at law and intends to settle in your neighbourhood. you will greatley oblidge me by throughing any Business in his way you conveniently can I think you will find him an active man in his profeshion. There has been many upes and downs in this neighbourhood since you left you would no doubt have heard that Charles Nixon was throughing out at Walbottle Collery by his partners some years ago he has little to depend on now but the profets of the ballast machine at Willington Quey wich I darsay is verey small many of his Familey has turned out verey badley he has been verey unfortunate in Famaley affairs. If, I have the pleasure of seeing you I shall give you a long list of occurences since you and I worked together at Newburn. Hawthorn is still at Walbattle I darsay you will well remember he was a great enamy to me but much more so after you left. I left Walbattle Collery soon also after you and has been verey prosperous in my concerns ever since I am now far above Hawthorn's reach. I am now concerned as Civil Engineer in different parts of the Kingdom. I have onley one son who I have brought up in my own profeshion he is now near 20 years of age I have had him educated in the first Schools and is now at Colledge in Edinbro' I have found a great want of education myself but fortune has made a mends for that want.

> I am dear sir yours truly
> Geo. Stephenson.

Killingworth Collery.

George had, indeed, raised himself thus early to be 'concerned as a civil engineer in different parts of the kingdom.' With a salary of £300 from the Stockton and Darlington Railway, with a rapidly increasing business, and with important accumulations, he found himself, in 1823, a made man. He could therefore well afford to defray the expenses of his son's visit to the university of Edinburgh.

Of that visit perhaps the most important result was the commencement of Robert Stephenson's friendship with Mr. George Parker Bidder, late President of the Institution of Civil Engineers. Mr. Bidder, who had already been for two years studying at the university, was immediately attracted to Robert Stephenson by the mildness of his disposition, and at the same time by his plain common-sense intellect. During the university term they were nearly inseparable, as in after life they fought their parliamentary battles side by side. To the close of his life Robert Stephenson's happiest days were spent in his friend Bidder's family circle.

With Robert Stephenson's return from Edinburgh to Killingworth, the period of his West Moor life may be regarded as closed. On receiving his formal appointment as engineer to the Stockton and Darlington line, George Stephenson left Long Benton, and Robert accompanied his father as assistant in the new undertaking.

The construction of the Stockton and Darlington line did not preclude George Stephenson and his son from making long journeys to various parts of the United Kingdom in the discharge of professional duties. In the September of 1823 they went to Ireland, from which country Robert wrote with his accustomed energy and confidence to Mr. Longridge.

Dublin : Sept. 10, 1823.

DEAR SIR, — We have just arrived at Paddy's Land 'in far Dublin city.' We left London on Monday, at half-past one o'clock, travelled all night, and reached Bristol the next morning, and expected to have got the steam packet to Cork, but we were disappointed on being informed that the Cork packet had broken her machinery a few days before, and was laid up for repair. We were therefore obliged to come on to Dublin, upwards of two hundred miles out of our way. We leave here this evening in the mail, and shall arrive at Cork to-morrow evening, where we shall probably remain a few days, and then make the best of our way into Shropshire. The concern we are going to at Cork was set fire to by the mob, where the disturbance has been for some time. We expect to reach home in the course of a fortnight. When we were in London my father called at Mr. Gordon's office, but found he had set off the preceding evening to the North. My father desires to be remembered to him with his sincere respects. We hope by this time we have got our fortunes made safe with the Lord of Carlisle's agents. We have some hopes of some orders for steam engines for South America, in the Columbian States. *This, however, depends on the success of Perkins's new engine.* My father and he have had a severe scold. Indeed the most of the birkies were embittered at my father's opinion of the engine. He one day stopped the engine by his hand, and when we called the next day Perkins had previously got the steam to such a pitch (equal 15 atmosphere) that it was impossible for one man to stop it, but by a little of my assistance, we succeeded in stopping it by laying hold of the fly-wheel. This engine he formerly called an 8 or 10 horse-power, but now only a 4. I am convinced, as well as my father, that Perkins knows nothing about the principle of steam engines.

I remain, dear Sir,

Yours sincerely,

ROBERT STEPHENSON.

P.S.—You shall hear from us at Cork.

The story of George Stephenson's practical criticism on the merits of Perkins' engine is well known.

From Cork, Robert Stephenson wrote to Mr. Longridge.

Cork: Sept. 16, 1823.

DEAR SIR,—We left Dublin on the evening of the day we wrote our last, for Cork, in the mail, and we were not a little alarmed, when it stopped at the post office, to see four large cavalry pistols and two blunderbusses handed up to the guard, who had also a sword hung by his side. I can assure you, my father's courage was daunted, though I don't suppose he will confess with it. We proceeded on, however, without being in the least disturbed, except, now and then having our feelings excited by the driver, or some of our fellow-passengers, relating, and at the same time pointing towards the situation, where some most barbarous murder had recently been committed. In one instance, a father, mother, and son had been murdered one evening or two before. As we passed along, everywhere distress seemed to be the prevailing feature of the country, and this to an incredible degree among the poor. Indeed, numbers of them appeared literally starving. We frequently have read accounts in the English newspapers of the distressed state of Ireland, but how far they fall short of conveying a just idea of it. With regard to the appearance of the cities Dublin and Cork, I must say the former falls far short of the description given of it by some Irishman in the steam packet, as we came over from England. I asked some of them if it was equal to Edinburgh, and they seemed insulted at the comparison, but I can now say they ought to have felt highly honoured. Dublin excels certainly in size and business, but as to scenery and beauty of building, it shrinks into insignificance.

We were very kindly received at the Dripsey Paper Works by Macnay's family, and have just finished our business with them for the present, and intend leaving Cork in the steam packet this day for Bristol. From there we shall make the best of our way to Shifnal in Shropshire, and our business there will probably detain us five or six days. A small boiler will be wanted to send to Ireland. You will receive the order by George Marshall, or some of our people, in a few days. I hope Mr. Birkinshaw will see the plates nicely cut, as we want it neatly finished.

The most valuable part of Robert Stephenson's education was, however, yet to come.

CHAPTER V.

PREPARATIONS FOR AMERICA.

(ÆTAT. 20–21.)

George Stephenson's Rupture with Mr. Losh — The Establishment of the Firm of R. Stephenson and Co. of Newcastle — The Colombian Mining Association — George Stephenson a Chief Agent for the Project — Robert Stephenson visited with renewed and aggravated Symptoms of Pulmonary Disease — Robert Stephenson proposed as Engineer to the 'Colombian Mining Association' — His Visits to Cornwall and other Places — Newcastle — The London Coffee House, Ludgate Hill — Robert Stephenson accepts the Post of Engineer-in-Chief to the Colombian Mining Association — In London — Preparations and Hard Work — 'Home, sweet Home' — Letter to 'the North' — Conduct of 'the Association' — Liverpool — Sails for South America.

IN forming his new connection at Darlington, George Stephenson made the acquaintance not only of Mr. Pease, but also of Mr. Michael Longridge of the Bedlington Iron Works, and the influential associates of both those gentlemen; and by his conduct towards them he gained their respect and confidence. Unfortunately, however, in acting honourably towards his new friends, he was compelled to give offence to an old patron. On being asked what rails he would recommend to be laid down on the Stockton and Darlington Railway, he frankly replied to the directors—'Gentlemen, I might put £500 into my pocket by getting you to buy my patent cast-iron rails. But I know them. Take my advice, and don't lay down

a single cast-iron rail.' Of course it was his paramount
duty to give this advice to his employers, but his con-
demnation of cast-iron rails, and recommendation of
malleable bars, not only kept £500 out of his own
pocket, but withheld the same sum from the purse of
his co-patentee and old employer, Mr. Losh. The latter,
not then believing in the relative inferiority of the cast-
iron rails which he and George Stephenson had patented
in 1816, was naturally irritated, and imprudently wrote
a letter to Mr. Pease reflecting on George's conduct in
violent and unjust terms. The contents of this epistle
were inconsiderately imparted by Mr. Pease to George
Stephenson; and the consequence was a stormy interview
between the latter and Mr. Losh, in which the capitalist
accused the engineer of ingratitude, and the engineer re-
torted on the capitalist a charge of self-interest and
cupidity. The consequence of this was, that the rupture
between the elder Stephenson and Messrs. Losh, Wilson,
and Bell was final; and George attached himself to another interest.

Whilst he was superintending the construction of the
Stockton and Darlington Railway, George Stephenson
induced Mr. Edward Pease, Mr. Richardson, and Mr.
Longridge, to join him in establishing the 'manufactory,'
now celebrated, wherever locomotive engines are used,
under the name of 'Robert Stephenson and Co.' It has
been already seen how he put Robert Stephenson's name
on the map as engineer of the Stockton and Darlington
line. In like manner, now that he was about to embark in
a great commercial speculation, he made his son the prominent engineer, as well as an actual partner, and was

pleased to keep himself in the background. The partnership was formed in 1823, and forthwith the ground was purchased on which the factory of 'Robert Stephenson and Co.' yet stands — an imposing and extensive mass of building, visible to travellers through smoke and fog, as the train bears them along the superior road of the High Level Bridge. The originators of the factory, interested deeply in the Stockton and Darlington Railway, were bent on supplying the new line with the steam locomotives, which their influence would cause to be adopted in preference to fixed engines. With the commencement of 1824 the factory was at work. George Stephenson, fully engaged with the Stockton and Darlington line, thirty or forty miles distant from Newcastle, could give but little personal care to the new factory. Robert Stephenson was, therefore, called upon to superintend its earliest operations. It was a trying position for a young man, only twenty years of age. To be so trusted was the grandest sort of education — but it was an education fitted only for an able man. He had to supervise the building operations, engage men, take orders, advise on contracts, draw plans, make estimates, keep the accounts, and in all matters, great or small, govern the young establishment on his own responsibility.

All this, however, was mere child's play compared with his next task.

A more fascinating scheme than that of the 'Colombian Mining Association' had not for years roused the imaginations of speculators. The proposal was to recommence working in Spanish America the gold and silver mines, which, it was averred, had been wrought with great profit before the Revolution. The cautiously expressed opinion of Humboldt, that such operations might lead to successful results, induced men of wealth and high reputation in the money market to support the project with their names and their gold. The first plan of the projectors was departed from in important particulars; and when the Company took form as a working power, its title was the 'Colombian Mining Association,' and the attention of the directors was concentrated on the mineral wealth of Colombia.

Amongst the most sanguine projectors of this speculation was Mr. Thomas Richardson, the founder of the famous discount house of Richardson, Overend, and Gurney. Mr. Richardson was an intimate friend and family connection of Mr. Pease of Darlington. He took shares in the Stockton and Darlington Railway, and became a partner in the firm of 'Robert Stephenson and Co., of Newcastle.' Frequently coming into contact with George Stephenson, he admired his soundness of judgement as much as he did his genius for mechanical contrivance, and consequently consulted him on the arrangements of the 'Colombian Mining Association.' Of course, steam-engines and iron goods would be required in abundance for effectually working the old mines; and Mr. Richardson calculated that his influence would obtain large orders for the house of 'Robert Stephenson and Co.' On George Stephenson, therefore, it eventually devolved to select miners, artisans, inspectors, and implements, and to make heavy shipments of iron and goods for America. Indeed, not only Mr. Richardson, but the general body of directors, relied on George's guidance in all the engineering part of their preliminary operations.

Although the earlier commissions were sent to his father, young Robert Stephenson had to attend to many of them ; and he did the work in such a manner that Mr. Richardson formed a yet higher opinion of his energy and capacity. Mr. Longridge, with whom George Stephenson had now, for more than three years, been in communication, also formed the highest estimate of Robert's abilities. Overtures were then made through Mr. Richardson to Robert Stephenson, sounding him whether he would like to accompany the expedition. The proposal put the young man in a fire of excitement. He was pining to get away from Newcastle. The threatening symptoms of pulmonary disease, which had from childhood made his friends anxious for him, seemed decidedly on the increase ; and in his secret heart he believed that the harsh winds of Newcastle would, before many years, lay him in a premature grave. In the warm luxurious atmosphere of Colombia, surrounded by the gorgeous beauties of animal and vegetable life, which had stirred Humboldt from his philosophic calm, he anticipated renewed vigour of mind and body. Moreover, the dreams of wealth, which had fascinated apparently cautious and practical men like Mr. Richardson, seemed to Robert Stephenson's young mind no visionary hopes, but realities beyond the reach of doubt. He argued, not unreasonably, the Spaniards, with imperfect appliances and a rude knowledge of their art, extracted from those mines vast revenues, and therefore greater wealth will flow to labourers aided by the latest inventions of science, and having a supply of skilled artisans.

It was true 'the works' had been scarcely established at Newcastle, and needed vigilant direction. But the principal object for which they had been started—the construction of locomotives—could not be attained until there was a public demand for the commodity; and even to Robert Stephenson, not less sanguine than his father as to the ultimate success of the locomotive, it seemed highly improbable that the demand would be either urgent or general for some years. At all events he might with advantage to his health and prospects go to South America for three years. George Stephenson did not at all like the proposal. Not even the annual salary of £500, with allowances for travelling expenses, could lessen his disapproval.

In the spring of 1824, Robert Stephenson, at the direction of the Colombian Association, went on a trip to Cornwall, accompanied by his uncle Robert (the father of the present Mr. George Robert Stephenson), and made a careful examination of the mining system of that country. The result of this trip was an elaborate report by the uncle and nephew on Cornish mining — its usages, implements, engines, and commercial organisation. Writing to his father from Oakhampton, Devonshire, March 5, 1824, Robert Stephenson said:—

As far as I have proceeded on my journey to the Cornish mines, I have every reason to think it will not be misspent time; for when one is travelling about, something new generally presents itself, and though it is perhaps not superior to some scheme of our own for the same purpose, it seldom fails to open a new channel of ideas, which may not unfrequently prove advantageous in the end. This I think is one of the chief benefits of leaving the fireside where the young imagination received its first impression.

In this same letter he speaks of having inspected the
Bristol steam-boats, and especially the 'George IV.,' in
which he and his father had crossed from Ireland in
the previous year.   He mentions also having been at
Swansea, where the engine for drawing coals, put up
by George Stephenson, was seen working admirably.
Speaking of the Neath Abbey Works, he observes :—

> When I was at Neath Abbey I had the pleasure of being
> introduced to Mr. Brunton the engineer: he is a very sensible
> man, but there is not one of them who understands the parallel
> motion thoroughly.   They seemed to doubt me when I told
> them I had never seen one mathematically true, not even in
> principle.

In the firm and self-reliant tone of this passage may be
seen the young man of twenty-one conscious of his power
to be a leader of others.

Returning to Newcastle, Robert Stephenson found that
he could not settle down to his work.   He wrote to his
father, begging him no longer to oppose his wish to go to
Colombia.

> But now (he wrote) let me beg of you not to say anything
> against my going out to America, for I have already ordered
> so many instruments that it would make me look extremely
> foolish to call off.   Even if I had not ordered any instruments,
> it seems as if we were all working one against another.   You
> must recollect I will only be away for a time; and in the mean
> time you could manage with the assistance of Mr. Longridge,
> who, together with John Nicholson, would take the whole of
> the business part off your hands.   And only consider what an
> opening it is for me as an entry into business; and I am in-
> formed by all who have been there that it is a very healthy
> country.   I must close this letter, expressing my hope that you
> will not go against me for this time.

Sorely against his will, George gave his consent; and

Robert Stephenson, once more going up to London, took up his quarters (April 27, 1824) at the London Coffee House, Ludgate Hill, and made his preparations for departure. It was a terribly wet season, and he walked about day after day in the flooded streets, soaked to the skin, buying implements and stores and engaging workmen. Nor did he confine his attention to the concerns of the Colombian Association. Already he was a man of mark, invited to the tables of wealthy merchants, and carried hither and thither to give his opinion on engineering questions relating to gas works, water works, and marine engines. He examined minutely Mr. Brown's 'vacuum engine,' which was making as great a stir as Perkins' machine did, until George Stephenson, by the simple application of muscular force, stopped the action of the pretty toy. The 'vacuum engine' Robert Stephenson significantly described in a letter to his father as 'extremely ingenious, but ——.' At the same time he busied himself in inventing, for a company of London merchants, a machine for stamping coin, which he hoped to see employed in the Colombian mint. The Messrs. Magnays had given him an order for a paper-drying machine. Whilst he was deciding how he should construct the machine for stamping coin and the drying machine, he visited the Mint and the 'Times' Newspaper Office; with which establishments he was so pleased that he wrote his father a long account of them.

The Magnays (he wrote) got me an introduction to the 'Times' printing office, where I was almost as much delighted as I was in the Mint. The facility with which they print is truly wonderful. They were working papers at the rate of 2,000 per hour, which they can hold for any length of time.

The mode they have of conveying the sheet of paper from one part of the machine to the other, is, I think, precisely what is wanted in the drying machine.

Hitherto Robert Stephenson's experience as a mining engineer had been principally confined to coal mines, whereas he was now about to search for the precious metals. That he might be possessed of all the requisite practical information, he took lessons of Richard Phillips, the Professor of Mineralogical Chemistry—the Colombian Company paying five guineas for each lesson. At the same time he was acquiring the Spanish language.

After staying for a short time at the London Coffee House, he removed to lodgings in No. 6 Finsbury Place South, and there remained till he left London. In 'the city' he underwent much disappointment. Arrangements which had been spoken of as completed had still to be begun. Heavy arrears of labour fell upon the young engineer, in respect of matters about which he ought to have had no trouble whatever. Even about his appointment—the salary and exact character of the position — there were difficulties ; and he had to haggle and insist before he could get any recognition whatever of his engagement with the Colombian Mining Company; and after all his agreement was not with the Company, but with the Company's agents, Messrs. Herring, Graham, and Powles, in their individual capacity. Thus after all Robert Stephenson sailed from England the agent of *the firm*, although he was to preside over the engineering affairs of *the Association*. All this augured ill for the state of affairs in South America.

During his protracted stay in London, whilst he was acquiring scientific information, purchasing stores, and vainly endeavouring to ascertain what his duties would be in South America, Robert Stephenson wrote to his friend, Mr. Longridge, in March, and again in April. The March letter was written at a time of great distraction and uncertainty, just after his return from Cornwall. The April letter was penned after a brief excursion in the country.

> Imperial Hotel, Covent Garden:
> March 9, 1824.

DEAR SIR, — Your letter the other day gave me pleasure in hearing you were going on (I suppose, of course, at Forth Street) pretty regularly. I wrote to my father this morning, but positively I durst not mention how long it would be before I should be able to reach once more the North. Indeed, I scarcely dare give it a thought myself. I saw Mr. Newburn yesterday, and he informed me it would at least be fourteen days before I could get my liberty. For heaven's sake don't mention this to my father. Joseph Pease will perhaps give him the information: it will, I know, make him extremely dissatisfied, but you know I cannot by any means avoid it. There are some new prospects here in agitation, which I look forward to with great satisfaction. It is the making of a road in Colombia. What a place London is for prospects! This new scheme of the road or railway is also connected with four silver mines at Mariquita. The road is projected between La Guayra and the city of Caraccas. You may find La Guayra on the coast, I believe, of the Gulf of Mexico. The climate, from Humboldt, is not quite so salubrious as that of Mexico. Mr. Powles is the head of the concern, and he assures me there is no one to meddle with us. We are to have all the machinery to make, and we are to construct the road in the most advisable way we may think, after making surveys and levellings.

Well might Robert Stephenson say, 'What a place London is for prospects!' He had come up to London to settle about going to South America as engineer of the Colombian Mining Association, and after all the

principal promoters of that association now proposed
to send him out on a distinct expedition to another
spot, although in the same quarter of the globe. Eventually, as it has already been stated, he went out as the
servant of Messrs. Graham, Herring, and Powles; and it
was his intention, when he had attended to their business,
to enter on the work of the Mining Company.

After many delays the agent of Messrs. Herring,
Graham, and Powles, and Engineer-in-chief of the
Colombian Mining Company, received orders to proceed
immediately to Falmouth, and there take ship to
Carthagena. The principal goods and the first lot of
miners had already quitted England, and the interpreter
to the expedition was already en route for Falmouth.
Obeying his instructions, Robert Stephenson had actually
mounted the Falmouth coach, and had loaded it with
extra luggage, to the amount of a £30 fare, when he
received orders to descend, to unload the coach, and to
start for Liverpool. Of course he complied.

On reaching Liverpool he wrote to his father (June 8,
1824), giving an account of his journey from town that
affords a striking picture of the troubles of ' the good
old coaching days.'

We have arrived safe in Liverpool, after an extremely
fatiguing journey. I never recollect in all my travels being so
terrified on a coach. I expected every moment for many miles
that we should be upset, and if such an accident had happened
we must have literally been crushed to pieces. We had 21 cwt.
of luggage to remove from London to Liverpool by coach. This
may serve to give you a faint idea of the undertaking. This
weight was sent in twice. The coach-top on which we came
was actually rent; all the springs, when we arrived at Liverpool,
were destitute of any elasticity, one of them absolutely broken

and the body of the coach resting on the framework, so that, in fact, we rattled into this town more like a stage-waggon than a light coach.

On June 12, George Stephenson arrived in Liverpool to bid his son farewell, and took an affecting leave of him on the 18th.

During his stay at Liverpool with his son, George Stephenson, by the hand of a friend, wrote the following characteristic and entertaining letter to Mr. Longridge :—

Liverpool: June 15, 1824.

DEAR SIR, — I arrived here on Saturday afternoon, and found Mr. Sanders, Robert, and Charles, waiting for me at the coach office. It gave me great pleasure to see Robert again before he sails. He expects to leave the country on Thursday next. We dined with Mr. Sanders on Saturday, and with Mr. Ellis yesterday. He had three men-servants waiting in the entrance-hall to show us to the drawing-room. There was a party to meet us, and kindly we were received. The dinner was very sumptuous, and the wine costly. We had claret, hock, champagne, and madeira, and all in great plenty; but no one took more than was proper. It is a good custom not to press people to take so much as does them harm. We dined at seven and left at twelve o'clock. Sanders and Ellis are magnificent fellows, and are very kind; Mrs. Sanders is a fine woman, and Mrs. Ellis very elegant. I believe she is niece to Sir James Graham, M.P.; I must say that we have been very kindly received by all parties. I am teased with invitations to dine with them, but each indulgence cannot be attended by me. What changes one sees !— this day in the highest life, and the next in a cottage — one day turtle soup and champagne, and the next bread and milk, or anything that one can catch. Liverpool is a splendid place—some of the streets are equal to London. The merchants are clever chaps, and perseverance is stamped upon every brow There is a Doctor Trail, a clever mineralogist, and some famous mathematicians that we have dined with. I was much satisfied to find that Robert could acquit himself so well amongst them.

He was much improved in expressing himself since I had seen him before; the poor fellow is in good spirits about going abroad, and I must make the best of it. It was singular good-fortune that brought us together at this time, but the weather is very bad; it has poured with rain for the last three days. To-day I am going over part of the line, but have not been able to commence yet. Robert will endeavour to write to you before he sails, and desires his kindest remembrance.

God bless you, Sir!

Believe me to remain

Yours sincerely,

G. S.

As soon as his father had said farewell, Robert Stephenson, before he went on board, wrote a hasty line, full of filial tenderness, to his mother, explaining that he had directed Messrs. Herring, Graham, and Powles to pay £300 per annum, i. e. three-fifths of his salary, to his father. For several years after their establishment 'the works' at Newcastle did not pay their expenses. George Stephenson's partners were far from sanguine as to their ultimate success, and George, confident as he was that they would prove a source of great wealth, was often pinched for ready money to meet his share of the capital required to feed them: Robert Stephenson knew this well, and did his utmost to meet the difficulty.

On the evening of that same June 18, on which he took leave of his father, Robert Stephenson wrote in his log-book :—

June 18, 1824.— Set sail from Liverpool in the 'Sir William Congreve,' at three o'clock in the afternoon: wind from the south-east, sea smooth, day beautiful; temperature of the air towards evening in the shade, 58°. Made some experiments with 'Register Thermometer' to ascertain the temperature of the sea at various depths, but failed on account of the velocity

of the vessel through the water not allowing the instrument to
sink. The temperature of the surface water appeared to be
54° at seven o'clock in the evening — this ascertained by
lifting a bucket of water on board and immediately immersing
the thermometer. This was considered as sufficiently accurate, as
the temperature could not sensibly change in the time occupied
by the experiment.

Pursuing the system commenced on that first lovely
evening at sea, Robert Stephenson jotted down in his
log-book the mutations of temperature and light, and
other natural phenomena, until on July 23, 1824, he
records :—

Early in the morning saw the Colombian coast, and at two
o'clock cast anchor opposite La Guayra ; observed with silence
the miserable appearance of the town. The hills behind the
town rise to a height that gives a degree of sublimity to the
scenery in the eyes of a stranger.

The voyage was at an end.

CHAPTER VI.

SOUTH AMERICA.

(ÆTAT. 20-24.)

La Guayra — Caraccas — Proposed Breakwater and Pier at La Guayra — Survey for Railroad between La Guayra and Caraccas — Santa Fé de Bogota — Mariquita — Life on the Magdalena — Explores the Country — Road between the Magdalena and the Mines — Santa Ana — Descriptions of Scenery — Arrival of the Cornish Miners — Insubordination of Miners — Friends, Pursuits, and Studies — Inclination and Duty — Disappointment of the Directors — Their Secretary.

LANDING in La Guayra on July 23, 1824, Robert Stephenson had to direct his attention to three important affairs and report thereon to Messrs. Herring, Graham, and Powles — the propriety of constructing a breakwater before the harbour of La Guayra, the cost and policy of building a pier for the same port, and the possibility of uniting La Guayra and Caraccas by a line of railway.

His reports on these three propositions were full and decisive. Having ascertained the characteristics of the harbour, the nature and declivity of the bottom of the shore, and the direction and force of the seas at different seasons, he pronounced that the construction of the breakwater would be a dangerous experiment.

A correct idea of the seas (he wrote) sometimes experienced in this port cannot well be conveyed by description. One circumstance, however, which may give some idea of their force is worthy of remark. It occurred during a storm last year, when a number of ships were wrecked. A large block of stone, upwards of two tons weight, measuring about eight feet long, four feet broad, and one foot thick, was thrown up by the waves four feet above the usual level of the sea, and such was the violence with which it was projected, that on its coming in contact with the other fragments of rocks on the shore, it was divided into two pieces, one of which now lies considerably out of the reach of ordinary seas. It is very remarkable that during the storm to which I have just now alluded, scarcely a breath of wind prevailed, while the sea raged with such violence as to drive every ship in the harbour from her anchors, and several were wrecked on the coast. The cause of this extraordinary phenomenon is yet unknown to us. It is not improbable that it was some branch of the Gulf Stream, modified by the conformation of the coast, the nature of the soundings, and many other circumstances combined, with which we are totally unacquainted.

Though he condemned the project of a breakwater, he advised the construction of a pier; and in support of this counsel he gave returns of the imports and exports of the harbour, the amount annually raised for wharfage of goods, and the insufficiency of the existing pier for the business of the port. The cost of such a pier as he advised (140 yards long and 24 feet wide at the top) would be £6,000, including the freight of workmen and of the necessary machinery to be sent out from England. The principal material of the structure would be the stone of the adjacent mountains, which could be conveyed by a short railroad to the site of the pier. In sinking the blocks of stone, he advised that care should be taken to 'give the pier a gradual slope on the seaward side, so that the waves might be completely broken, and consequently their force almost totally extinguished, before reaching the body of the pier.'

When he came to consider the third and most important of the three propositions — the construction of a railway between La Guayra and Caraccas — the advantages likely to follow from the undertaking, and the natural obstacles to the work, caused him much anxious thought. The ground was very different from any on which he had ever seen rails laid. Mounting a mule, he surveyed the road between the two towns, and found it 'a wonderful example of human industry — not of human skill.' The ascents and descents were so precipitous that he wondered how his brute contrived to keep on its legs.

To give you an idea (he wrote to his father) of the trouble I have already had in seeking for a new road, and the trouble I shall yet have, would be an impossibility. You may attempt to conceive it by imagining to yourself a country, the whole surface of which, as far as the eye can reach, is thickly set with hills, several thousand feet high, from six to eight times as large as Brusselton Hill. There is a valley, however, which extends the whole way nearly between La Guayra and Caraccas, up which I think is the only situation we could get a good road; but even in this valley there are hills as high as Brusselton. I dare not attempt any tunnelling, because the first earthquake — and there is no knowing how soon it may come — would close it up, or at all events render it useless. This circumstance, you will agree with me, puts tunnelling out of the question. And to make any very extensive excavations with high sides would prove equally fatal on the occurrence of an earthquake.

As he rode up the valley of Caraccas, with mountains on either side, he saw that to put down a colliery tramway in Northumberland, and to lead a line of rails through such a ravine, were widely different tasks.

Having thoroughly examined the proposed line, he came to the conclusion that, with everything in his favour, he could lay down the contemplated railway for about £160,000. The great risks, however, that would attend the operations made him see that speculators would not embark their money in the affair unless there was a probability of at least a 10 per cent. dividend. The annual goods traffic between La Guayra and Caraccas did not amount to more than 5,571 tons. Therefore, if the road were made and opened, Robert Stephenson could not see his way to more than £14,180 profit on each year's transactions — an annual revenue that would only pay 10 per cent. on a capital of £140,000. Against the probability that the estimated £160,000 would be exceeded, he put the fact, that large quantities of goods, of which he could get no returns, were annually conveyed between the two towns. Again, traffic would be augmented by the stimulus which a railway would give to commerce and agriculture. The question admitted of much debate; but Robert Stephenson, with that prudence which preserved him in after life from brilliant indiscretions, concluded his report with saying : ' I think it would not be prudent at the present moment to commence the speculation.'

Whilst he was thus engaged at La Guayra and Caraccas, the miners with whom he had come out from Liverpool went on to Carthagena, and thence along the River Magdalena.

As soon as he could get away from Caraccas, he mounted his mule, and, accompanied by a black servant and by Mr. Walker, the interpreter to the expedition, proceeded across the country to Santa Fé de Bogota.

The journey was one of fatigue and peril. Cut-throats and ruffians were numerous in the country; but being well armed, Robert Stephenson went his way unconcerned. He was very anxious to reach Mariquita, near which place the principal mines of the Colombian Association were situated; but the nature of his duties forced him to travel slowly. Messrs. Herring, Graham, and Powles had instructed him to examine the mineralogical characteristics of the country in every direction; and in spite of the care he took to conceal the object of his journey, it soon leaked out that he was the engineer of a new mining company, and daily he was accosted by strangers, ready to mislead him with false information. More than once he was induced, by misrepresentations, to ride a hundred miles after a mare's nest. On one occasion he spent several days in following a guide, who promised to bring him to a fissure in a rock abounding with quicksilver. On reaching the spot the quicksilver was there; and he could not account for its presence, till a former governor of the district told him that a bullock-wagon loaded with quicksilver had, some years before, been upset in that spot. On reaching Bogota, however, he wrote to his father on January 19, 1825, expressing great confidence in the mineral wealth of the country.

Having reached Mariquita, he forthwith proceeded to examine the mines of the surrounding country. On every side he found workings; some of which had evidently been deserted because they offered no prospect of gain, whilst the appearance of the others induced a belief that scarcity of labour and capital, during the revolutionary struggles of the country, had been the sole reason for leaving them.

Mariquita was a spectacle at once imposing and mournful. Two-thirds of its habitations were in ruins. Heaps of rubbish covered sites formerly occupied by palaces. Of the public buildings, none were in a state of repair, except five churches. The convents were untenanted, and in dilapidation. Such was the havoc wrought by earthquakes, stagnation of trade, and disturbed politics, that of the population of 20,000 who had once inhabited the city, only 450 persons remained to see the entrance of Robert Stephenson, and wonder what had brought him to their ill-starred city.

Honda being the extreme point of the Magdalena navigable by craft coming from Carthagena, he hastened to inspect the route between the river port and the city in the interior, to which his men with their ponderous implements and machinery were advancing. The distance between Honda and Mariquita is about twelve miles, and the features of the country can be briefly stated. On leaving Honda the road is for a short distance precipitous, after which it rises gently for about two miles to an extensive breadth of table-land, beautifully covered with delicate grasses, and studded with groups of trees, some of which are in blossom at all seasons of the year. At points this magnificent plain is bounded by small isolated ridges of alluvial rocks. Some of these rocks are almost perpendicular from their bases up to their irregularly serrated peaks. Onwards the scenery is of increasing loveliness, and before Mariquita is reached, the route passes through groves of palm and coco, orange, cinnamon, and almond trees, pines and mangoes.

On the whole, the roads from the Magdalena to the mines in the immediate vicinity of Mariquita (the mines of Santa

Ana, La Manta, San Juan, and El Christo de Laxas) were good — that is to say, good for Spanish America. A moderate amount of labour would have rendered them passable for wheeled carriages, except at certain points where it was clear that wheels could never run. In these precipitous portions of the route, which mules took two hours to cross, Robert Stephenson saw at a glance difficulties of which he had not been forewarned, and for which he consequently was unprovided. The heavier portion of the machinery could not be moved across country except on wheeled carriages.

In due course the first party of miners arrived, but they had to leave the greater part of their machinery on the banks of the Magdalena, and proceed to the mines with only the lighter implements, which could be packed upon the backs of mules. Of course an urgent request was despatched to London that other machinery might be sent out, so constructed, that each large machine could be taken to pieces, small enough for transport on mules. But before this message reached the directors, they had shipped off from Newcastle a large quantity of iron goods, which, on being thrown upon shore by the peons at Honda, remained, and to this day probably remain, useless and cased with rust. Robert Stephenson, however, did not lose heart. Taking his men, and the few implements which they could carry with them, he hastened to the mines, reopened them, explored their workings, and commenced working for ore.

The best mines, of which the Association had obtained leases from the Colombian Government, were those of St. Ana and La Manta, adjacent to the village of St. Ana. The distance between Mariquita and St. Ana is about twelve miles; but those twelve miles comprised the worst portions of the way from the river. After leaving Mariquita, the miners had to traverse a plain for six miles, when they entered on a broken tract watered by two rivers, which it was necessary to ford. The next six miles lay up the ides of mountains. Often the way ran over bare rocks, through narrow passages worn by the floods of the wet season, and down declivities so nearly perpendicular that no beast of burden, except a mule, could descend them. Standing on an eastern slope of the Andes, the village of Santa Ana (containing when the miners first reached it about nine cottages) afforded a grateful contrast to the desolate grandeur of the city in the plain. Instead of the intense heat of the valley beneath, its temperature was about 75° in the shade, and during the night 6° or 8° lower. A breeze played through the trees; and the soil, rich as the mould of an artificial garden, yielded fruit and vegetables in abundance.

On all sides (Robert Stephenson wrote to his stepmother) is an immense forest of fine trees, which are always green, no winter being known in these climates. The leaves are always gradually falling, but they are immediately succeeded by fresh green leaves. The ground descends suddenly from the front of our house for above a mile, in which small distance the fall is no less than 800 feet. From the bottom of this descent, the ground rises rapidly to the height of 1,000 feet, forming a mountain ridge which is covered to the very summit with strong trees that are always green. Beyond this small ridge of hills rise others still higher and higher, until their tops are covered with everlasting snows, and where not a spot of vegetation is to be seen, all being white with snow and ice.

A grander panorama than that enormous ravine, walled by forests, and crowned with peaks of gleaming whiteness,

cannot be conceived.   Clothing the curves of the interior
hills were tree-ferns and magnolias, groves of bamboo,
acacias, palms, and cedars.   Another picturesque feature
added charm to the landscape.   Fed by the gradual dissolution of distant snow, a river ran from the cool heights
into the hot air of the valley.   By tranquil pools pelicans
watched for their prey, and overhead, in the branches,
parrots and mocking-birds, monkeys and macaws, gave
colour and animation to the picture.   Flashing with
metallic lustre humming-birds darted from flower to
flower, disturbing the clouds of butterflies which floated
through the luxurious atmosphere.

Amidst such scenery Robert Stephenson spent more
than two years, endeavouring with inadequate means to
cope with gigantic difficulties, and suffering under those
petty troubles which are more vexatious than greater
miseries.

In the immediate vicinity of Santa Ana, the mountain-
river, falling over ledges of granite, had worn deep basins
in the rock.   One of these tarns Robert Stephenson selected for a swimming bath.   The granite sides of the
pit being almost perpendicular, bathers could not walk
gradually into the deep water.   In the centre, however,
was fixed a flat block of stone, the top of which was
about thirty-six inches below the surface of the water,
the distance between the bank and the stone being at one
point not more than three feet.   Bathers who could not
swim used to jump from the side to this natural table.
Unfortunately a sudden fall of rain caused a torrent
of water to raise this ponderous mass of stone, and
bear it downwards to the plain.   A few days later, a
gentleman attached to the mining expedition, who was

unable to swim, went to the tarn. Having leaped from
the point, where he expected to alight on the block, the
bather in another instant was struggling in the pool.
Fortunately Robert Stephenson, who was an expert
swimmer, came up just in time to plunge into the basin,
and catching the sinking man by the back of his neck,
conveyed him safe to shore.

It was not till the end of October, 1825, that miners
had been collected in sufficient numbers to commence
great operations. In that month a strong staff of Cornish miners made their appearance, and with them for
a time Robert Stephenson's troubles greatly increased.
Proper care had not been taken to select sober and
steady men. It was right that English workmen engaged
to encounter the perils of a South American climate
should be well paid, but the terms on which these
miners had been hired were far too high. Insolent
from prosperity, and demoralised by the long-continued
idleness of the voyage, they no sooner entered Honda
than they roused the indignation of the inhabitants by
excesses which outraged even South American morals.
Before Robert Stephenson made the acquaintance of
the men, he received a formal and angry remonstrance
from the Governor of Honda with regard to their conduct. The only thing to be done was to get them to
work with all speed.

I have no idea, (wrote Robert Stephenson from Mariquita to
Mr. Illingworth, the commercial manager at Bogota,) of letting
them linger out another week without some work being done.
Indeed, some of them are anxious to get on with something.
Many of them, however, are ungovernable. I dread the
management of them. They have already commenced to drink
in the most outrageous manner. Their behaviour in Honda

has, I am afraid, incurred for ever the displeasure of the
Governor, at all events so far as induces me to despair of being
able to calculate upon his friendly cooperation in any of our
future proceedings.   I hope when they are once quietly settled
at Santa Ana and the works regularly advancing, that some
improvement may take place.   To accomplish this, I propose
residing at Santa Ana with them for awhile.

There was reason for uneasiness.   Robert Stephenson
spoke firmly to the men, but he saw that his language,
though moderate and judicious, merely roused their re-
sentment.   Scarcely a day passed without some petty exhibition of disrespect and hostility ; and though in Santa
Ana they had fewer opportunities for gross licentiousness,
they could not be weaned at once from habitual drunkenness and indolence.   The supervisors or ' captains,' as
they were called, according to the custom of the Cornish
miners, were the most mutinous.   Mere workmen, and
altogether ignorant of the science of their vocation, they
were incredulous that any man could understand mining
operations who had not risen from the lowest employ-
ments connected with them.   In the Northumbrian coal
field, a distich popular a generation since runs —

> Trapper, trammer, hewer,
> Under, overman, and then viewer.

The Cornish ' captains ' in like manner were strongly in
favour of promotion from the ranks, and were reluctant
to obey the orders of a mere lad, and, what was worse still,
a north-country lad.   Their insolence was fostered by the
ludicrous respect paid to ' the captains ' by the natives,
both Spaniards and Indians, who, misled by the title, regarded them as superior to the young engineer-in-chief.
The 'captains' themselves immediately saw their advantage

—and in their drunkenness told both the workmen and the native population that Robert Stephenson was merely a clerk, sent out to pay them their wages and see that the expedition did not fail from want of funds.

Quitting Mariquita, where the rumbling of earthquakes had not allowed him many nights of unbroken rest, Robert Stephenson took up his residence on the mountains, the curate of Santa Ana putting a cottage at his disposal. A few weeks passed on, and there were alarming symptoms of a general mutiny of the workmen against his authority. A new arrangement of the men at the different mines was the occasion of open revolt. One night early in December, the most dangerous and reckless of the Cornish party assembled in an apartment of the curate's cottage. Wearied with a long day's work, Robert Stephenson had retired to rest in the next room, and was roused from his first slumber by the uproar of the rascals, who, mad with liquor, yelled out their determination not to obey a beardless boy. For more than an hour he lay on his bed listening to the riot —fearful that the disturbance might lead to bloodshed, and prudently anxious to avoid personal collision with the drunken rabble. Of course he knew that their insolent speeches were intended for his ears, yet he remained quiet. He was alone — his opponents were many. If the difficulty became an affair of blows, the weight of evidence would be all against him; and even if he were killed, he would be believed to have provoked the conflict by his own rashness. But when the insurgents proposed that the 'clerk' should forthwith be taught his proper place, he rightly judged that it would not do for him to remain longer in his private room when his presence might still the storm, and could not aggravate it. Rising, therefore, from his bed, he walked into the midst of the rioters — unarmed, and with no more clothing on him than his trousers and shirt.

At his first appearance there was a low murmur, followed by a deep silence. Taking up his place in the middle of the room, he drew himself up and calmly surveyed them. Silence having had its effect, he said quietly, ' It won't do for us to fight to-night. It wouldn't be fair ; for you are drunk, and I am sober. We had better wait till to-morrow. So the best thing you can do is to break up this meeting, and go away quietly.'

Cowed by his coolness, the men made no reply. For a minute they were silent, and turned their eyes on the ground ; and then, rising from their seats, they stumbled out of the room into the open air, to surround the cottage and pass two or three hours in shouting, ' One and all! —one and all! ' thereby declaring that they were one and all determined on revolt. Thus far master of the position, Robert Stephenson lit a cigar, and, sitting down in the room, allowed the tipsy scoundrels to see him through the open door calmly smoking.

The riot being renewed on a subsequent night, he left his cottage, and, accompanied by two friends, found refuge in the house of a native.

It appears remarkable (wrote Robert Stephenson to Mr. Illingworth, December 8, 1825) that having been all my life accustomed to deal with miners, and having had a body of them under my control, and I may say in my employ, that I should now find it difficult to contribute to their comfort and welfare. They plainly tell me that I am obnoxious to them, because I was not born in Cornwall; and although they are perfectly aware that I have visited some of the principal mines in that

county, and examined the various processes on the spot, yet
they tell me that it is impossible for a north-countryman to
know anything about mining.

Fortunately, Robert Stephenson had a cordial ally in
Mr. Illingworth at Bogota, who lost no time in sending
word that Robert Stephenson was the head of the expedition, and that the men from high to low were to obey
him, and him alone.   And in due course these representations were rendered yet more emphatic by letters from
the Board of Directors in London.

When a better feeling had been established between
the miners and himself, Robert Stephenson encouraged
them to spend their evenings in athletic sports.   In cast-
ing quoits, lifting anvils, reaching beams suspended by
cords, and throwing the hammer, he had few equals ;
and by displaying his prowess in these and similar sports,
he gradually gained the respect and affection of his men ;
but he was unable to work a complete reformation in
their habits.   To the last he could never get from any
man more than half a day's work each day, and he always
had nearly a third of his hundred and sixty subordinates
disabled by drink.

Having moved from Mariquita to Santa Ana, he
had a cottage built for his own habitation.   It contained
two rooms, the outer and inner walls being composed
of flattened bamboo, and the ceilings of smooth reeds,
palm-leaves being used for the roof.   The entire framework was tied together with cords of the tough and
pliant bijuco.   In this cottage, commanding a view of
the ravine, he was so fortunate as to have congenial
society.   Visitors came from Bogota and Mariquita, and
for weeks together he had with him M. Boussingault and

Dr. Roullin. The former was an accomplished chemist and geologist; and the latter had been invited by the Government to become Professor of Mathematics in an University which it was proposed to establish in the new republic. Under their guidance Robert Stephenson studied with system and accuracy the higher branches of mathematics, and various departments of natural science. Occasionally he made excursions to Bogota and Mariquita, to attend the horse-races or the balls; but such trips were only occasional relaxations, after weeks of work and study at Santa Ana. At this time, also, he took especial pains to rub off the remains of that provincial roughness which had marked him in boyhood. With characteristic simplicity he begged the few English gentlemen of his acquaintance to correct him whenever he used the diction, idioms, or intonations of north-country dialect. Knowing the disposition with which they had to deal, his friends took him at his word; and though at first their criticisms were frequent and far from pleasant, they never produced in him even momentary irritation. In one of his letters to his mother at this period he speaks of himself as dividing his time 'between eating and study.' In study he was perhaps intemperate, but in his diet he was habitually sparing and moderate. Occasionally he took wine and spirits, but his usual drink was water. He smoked regularly, but not immoderately.

To have a complete picture of Robert Stephenson's South American life, the reader must remember his strong love of animals, and imagine the bamboo cottage of the Andes peopled with four or five monkeys, as many parrots, and a magnificent mule named 'Hurry,' who, as soon as his master's dinner-hour arrived, used to enter the sitting-room, and patiently wait beside the table until he was presented with a loaf of bread.

Whilst he was thus living in his mountain-home he received on the whole but few letters from England. During the first twelve months, indeed, of his absence from his native land, he heard frequently from his father, as also from Mr. Edward Pease, Mr. Joseph Pease, Mr. Richardson, Mr. Longridge, Mr. John Dixon, Mr. Edward Storey, worthy Anthony Wigham of Killingworth, and Mr. Nicholas Wood; but as time went on, these correspondents became remiss, and Robert Stephenson learnt what grief it is to pine in a foreign land for one's own country, and at the same time to feel neglected by those at home. During the last twelve months of his stay in Colombia he did not hear once from home, either through the miscarriage of letters or the neglect of his father and stepmother to write. In a letter to Mrs. George Stephenson in the June of 1826, he observes, with a burst of that strong affection which inspired him to the last : —

My dear father's letter, which I received a few days ago, was an affectionate one, and when he spoke of his head getting grey and finding himself descending the hill of life, I could not refrain from giving way to feelings which overpowered me, and prevented me from reading on. Some, had they seen me, would perhaps call me childish : but I would tell them such feelings and reflections as crossed me at that moment are unknown to them. They are unacquainted with the love and affection due to attentive parents, which in me seems to have become more acute, as the distance and period of my absence have increased.

The longer he remained in South America the more painful was his position. A very brief acquaintance with the country satisfied him he was at the head of an enterprise projected by visionary speculators, who had no real knowledge of its difficulties. The letters which he received from England during the first year of his absence, showed that the unsoundness of the scheme was known to the best judges of such matters in London. It is not agreeable to be tied to a losing concern. He felt that no credit could come to him from his connection with the Colombian Mining Association, and he would gladly have ended it. This feeling was strengthened by his English correspondents. His partners in the concern at Newcastle begged him to return to look after the affairs of the factory, which were suffering by his absence. They represented to him that he had no legal agreement with the Company, and that Messrs. Herring, Graham, and Powles would not disapprove his immediate return.

But Robert Stephenson felt that he was bound to stay at the mines. It was true the Company had not a hold upon him in law, but it had in *honour*; and he resolved to remain, at any cost, till the stipulated three years had expired, or until he had obtained formal permission from the directors to leave his post.

The following letter, written to Mr. Longridge at the close of 1825, when he had hopes of honourable liberation from his distasteful engagement before the expiration of the three years, shows his state of mind : —

Mariquita : December 15, 1825.

MY DEAR SIR,— About a fortnight ago I received your kind letter, dated July 21, 1825. I was glad to learn your family was in good health, to whom I beg to be remembered in the kindest manner, as well as to my other friends in your part of the world. Your account of affairs in England was to me exceedingly interesting, particularly that part respecting the progress of the railway undertakings. The failure of the

Liverpool and Manchester Act, I fear, will retard much this kind
of speculation; but it is clear that they will eventually succeed,
and I still anticipate with confidence the arrival of a time when
we shall see some of the celebrated canals filled up. It is to be
regretted that my father placed the conducting of the levelling
under the care of young men without experience. Simple as
the process of levelling may appear, it is one of those things
that requires care and dexterity in its performance. Your
advice regarding my leaving this country, should my agreement
be transferred to the Colombian Association, I refrained from
following, principally from what Mr. Richardson said in his
letter, contained in the same sheet with yours, in which he
requested me not to leave the country without the consent of
my employers. This I was inclined to think was the most
advisable, especially as I have already been so long from
England, and that the stay of a few months longer might
secure me their interest on my return, and I still entertain
hopes of being able to leave this country previous to the
expiration of three years, as the agents in Bogota have recently
represented to the Board of Directors the assistance that I might
be to them in England in arranging such machinery as may be
required in this country. What they have sent out is a pretty
good specimen of the ideas they have of the difficulties to be
encountered in the conveyance of heavy materials. If Mexico
presents as many obstacles, and of equal magnitude, as Colombia,
I can say at once that a great number of the steam-engines that
were being made when I left may as well be made use of at home.

Since I wrote to you last about the Isthmus of Darien,
things have taken a turn. Messrs. Herring & Co. appear to
have relinquished, in a great measure, the idea of embarking
largely in making roads, and in consequence have raised a
private association, consisting of a few of the most respectable
houses in London, who have made such propositions to the
Colombian Government as seem to leave little doubt but they
will succeed in obtaining the privilege. Their wish is not so
much to retain the road, after it is made, altogether to themselves, as to lend the Government money and supply them with
English engineers under a certain interest, and afterwards to
share with the Government a proportion of the profits arising
from the road. These propositions display liberality, and are

of such a nature as, in my opinion, will induce the acceptance of
them.   This arrangement put an end to those that · had been
entered into by the agents in Bogota, and consequently renders
it uncertain whether I shall have to go or not.   For the same
reason, I suppose, the models that I wrote you about are lost
sight of.   At all events, I shall visit the isthmus in order to get
local information which may be of use to me in England, as I
feel quite satisfied that the scheme will go on.   We have heard
many objections urged against the project, such as the difficulty
of procuring European workmen in sufficient numbers, and more
especially the nature of the climate, which is said to be extremely bad, from the excessive and continual humidity which
reigns more or less throughout the whole year, and gives rise to
fever and ague.   Much doubt, however, exists on this score.   In
obtaining the privilege for sending steam machinery to the
country for the use of the road, I fear some obstacles have
arisen since I wrote you.   Congress, I believe, has thrown out
some hints that more attention would hereafter be paid to
granting monopolies of that description.   I have had a good
deal of conversation with the house in Bogota.   They seem to
think it better to mention it to Mr. Powles.   I see no advantage
in that; but I shall make such arrangements with R. S. Illingworth, the representative of the house, that, if nothing should
be done before I leave, a correspondence may exist between us.
I have had so much to do lately that I have not been able to
pay any attention to this matter.

I have my health just now very well, though I cannot say am
so strong as when I left England.   The tropical climates are far
from being so unhealthy as is generally supposed by those in
northern latitudes.   The rainy season is the only objectionable
part.   It occurs twice in one year.   The first season of rain at
Mariquita commences about the middle of March and continues
till the middle of May.   The second commences near the 11th
and 12th of October, and is just now terminating.   The remaining parts of the year are dry and hot, though not unhealthy.
Thermometer hot: in the morning 79° or 80°, at mid-day 82° to
84°.   During the rainy season it is 2° or 3° lower.   I have
once seen the thermometer as low as 73°, when I found it
uncomfortably chilly.   And at this moment it stands at 82°, and
not the least sign of perspiration about me, though I have been

walking. It is extraordinary how soon the human body becomes inured to high temperatures, without suffering much inconvenience. We have now got a steam-boat in action on the river Magdalena, being the second experiment; but the boat they have built last has the same fault as the first one—that is, drawing too much water. Much money has been spent in this speculation, chiefly from bad management. The engines are from the United States, where I have heard they have the finest steam-boats in the world; and as the communication from Carthagena to that country is frequent, I have some intention of seeing their steam machinery. It is the best way home, a regular packet being established between New York and Liverpool.

I hope soon to be able to give you some more certain details respecting my route home, as I fully expect from what has been said to the Board that I shall be liberated. I wrote to my father and mother about three weeks ago. I hope they have received my letter safe; but much uncertainty is connected with the forwarding of letters here. The post-office regulations are bad. The last letter that Mr. Pease wrote me came to hand open, from having been stuck to others by the melting of the sealing-wax, which almost invariably melts in these climates. Wafers are much preferable.

My kind love to my father and mother, and believe me,

My dear Sir,

Yours most sincerely,

ROBERT STEPHENSON.

P.S.—May I beg the favour of your attending to the payment of my yearly subscription to the Lit. and Phil. Society?* I rather suspect it has been neglected.

Michael Longridge, Esq.
    Bedlington Iron Works, near Morpeth, Northumberland.

So he remained, doing his best, and fighting with great difficulties. The amount of work he performed in

---

* The Literary and Philosophical Society of Newcastle, from which both the Stephensons derived so much benefit during their residence at Killingworth, and of which Robert Stephenson ultimately became chief benefactor.

the service of his employers was very great. He explored the country far and near; made assays of specimens of ore; wrote reams of letters and reports, many of which, besides being unexceptionable as business statements, have considerable literary merits; drew out a sketch for an efficient administration of mines; and in every way strove to earn and save money for the Association.

All these exertions met with no proper response in London. Instead of supplying him with the machinery for which he had written, the Directors sent out fresh cargoes of costly and ponderous apparatus, which could no more be conveyed over bridgeless rivers, and up mountain passes, than they could be wafted from the earth's surface to another planet; and to add to his chagrin, the projectors wrote to him, complaining that he had not already sent home a freight of silver. Some ignorant and self-sufficient persons reported to him the careless speeches and votes of the directors in the most offensive form. In answer to a statement in one of Robert Stephenson's reports, that the operations at Santa Ana might be accelerated if they had either steam, or water power wherewith to work certain machinery, one of the worthy officials reprimanded the engineer for not availing himself of such a noble river as the Magdalena. Of course he could only laugh at a proposition to turn the Magdalena up to the Andes. But when the Secretary undertook to criticise the investigations of M. Boussingault, the geologist and chemist employed by the Company, and presumed to sneer at the 'theoretical services' of the man of science, Robert Stephenson became indignant. 'These men,' he wrote, 'prate about the superiority of practical men over scientific men, being themselves neither the one nor the other.'

In his comments on M. Boussingault's proceedings, however, the London Secretary caused as much amusement as anger. In his report, the French *savant* had mentioned the advisability of using ' chiens' in the mines. On this information, the Secretary condemned in the strongest terms the cruelty of employing dogs as beasts of burden. In his next homeward despatch Robert Stephenson took an opportunity to inform the zealous protector of the canine race that the word *chien* in French, and *Hund* in German, was a mining term, signifying a kind of carriage with four wheels, which was not known in England by the name of *dog*, but by *tram* ; and that in the north of England a somewhat similar sort of carriage was known as a rolley.

CHAPTER VII.

FROM SOUTH AMERICA TO NEWCASTLE.

(ÆTAT. 23–24.)

Leaves Santa Ana — Goes up to Carthagena — Encounters
Trevithick — Trevithick's Peculiarities — Sails for New York — Becalmed amongst the Islands — Terrible Gales in the open Sea —
Two Wrecks—Cannibalism — Shipwrecked off New York — Strange
Conduct of a Mate—Is made a Master Mason—Pedestrian Excursion
to Montreal — Remarkable Conversation on the Banks of the St.
Lawrence — Returns to New York — Arrives at Liverpool — Meeting
with his Father — Goes up to London and sees the Directors of
the Colombian Mining Association — Trip to Brussels — Return
to Newcastle — Liverpool.

ROBERT STEPHENSON was aware that his prolonged
sojourn in America was highly prejudicial to his
interests.   Mr. Longridge, who during his absence had
undertaken the active management of the affairs of
' Robert Stephenson and Co. of Newcastle,' wrote urgent
entreaties for his return home.   His heart told him how
much his father needed him.   He knew, too, that all his
most influential friends — Mr. Richardson, Mr. Pease, and
other capitalists to whom he looked for countenance —
were of opinion that he might with propriety consult his
own advantage, in deciding whether he should quit, or
keep at his post.   His *word*, however, was given; and
he kept it.

At length the time came when he could honourably
start homewards : and as he looked back on the previous

three years he was not altogether dissatisfied with their results. From December 30, 1824, to December 31, 1827, the entire expenditure of the Colombian Mining Association had been little short of £200,000. A large. portion of this sum had been wasted by maladministration in London, but the great operations carried on with the remainder had been directed by him — a mere boy between twenty-one and twenty-four years of age. And in everything for which he individually could be held accountable the expedition had been successful. Had he worked the mines, as the Spaniards worked them, with the cheap labour of slaves, they would have yielded him as much profit as preceding engineers had extracted from them. As it was, on bidding official farewell to the directors, he was in a position to tell them that their property, under economical management and with the agency of proper machinery, could be made to pay them a handsome, though not an enormous, dividend.

In the July of 1827, Robert Stephenson wrote his last South American letter to Mr. Longridge.

July 16, 1827.

MY DEAR SIR, — The period of my departure from this place has at last really and truly arrived, though not longer than a month or two ago I was despairing of being able to get away without incurring the displeasure of the Board of Directors, as they wrote to the principal agent at Bogota, expressing an earnest wish that I would remain in St. Ana, notwithstanding my agreement having terminated, until the arrival of a new superintendent, whom they say they found great difficulty in procuring. Just about the same time I received a letter from Mr. Richardson, in which he states that the factory was far from being in a good condition, and that unless I returned promptly to England it would not improbably be abandoned. He further stated that the Board had not met with a person to succeed me; but notwithstanding this, he supposed I would leave at the expiration of my agreement. This induced me immediately to advise the agents in Bogota of my intention to leave with all convenient despatch, and of my hope that they would make such arrangements as might seem most expedient to them, respecting the filling up my situation. In answer to my letter, they determined upon coming down from Bogota to St. Ana, and attending the establishment themselves up to the arrival of another person from England. In pursuance of this resolution, Mr. Illingworth is now in this place, and it is my intention to leave on the 24th or 25th of the present month. By the 30th I shall have procured a boat at Honda for my passage to the coast. At present it is my intention to proceed direct to Carthagena, and I still have an itching to visit the Isthmus of Panama, so that I may know something about the possibility, or impossibility of forming a communication between the two seas; though the very short time that I can stay there will evidently prevent me getting more than a very general idea of such a scheme. From the information I have gathered from one or two gentlemen who have visited that coast, it would appear most judicious to proceed from Carthagena to Chagres by sea, and from the latter place to pass by the main road to Panama, on the Pacific—these being the situations between which a communication is most likely to be effected. It is extraordinary that the recent proposals which were made by British capitalists for undertaking this scheme to the Colombian Government did not excite more interest. When they were brought before Congress, they scarcely elicited a consideration; at least nothing was said, or done which the importance of the subject demanded. Some individuals of power connected with the Government were weak enough to imagine that a free communication between the Pacific and Atlantic Oceans would be productive of serious inconvenience to Colombia. Upon what grounds such an opinion was founded I am not well informed; but there can be no doubt but that interested views of this kind will in time fall to the ground, especially when civilisation has made more advances, and a more intimate intercourse between the inhabitants of the east and west parts of this continent shall be rendered almost, if not absolutely, necessary. From what I have seen of this republic, I feel thoroughly convinced that inland communication will ever remain imperfect—nay, probably little better than it now is.   Produce in the interior cannot possibly be conveyed to the coast, and thence exported to foreign markets, with profitable results; cultivation will consequently always be confined to the provinces bordered by the sea; I mean, of course, for such articles as are to be exported.   Whatever is yielded by the interior will be consumed at home.

If, therefore, a connection between the east and west populations of this continent is cut off by the natural difficulties presented by the surface, it seems reasonable to conclude that an opening by the isthmus to admit of conveyance by water will become indispensable.   This is only contemplating the advantage which such an undertaking would be to Colombia and the other South American powers.   But how the magnificence of such a work augments in our ideas when we consider the advantages which would arise from it — how it would influence commerce in every quarter of the earth !   The grounds of the proposal made by a number of the most respectable merchants' houses in London, for undertaking the examination and execution of a road, or canal across the isthmus, were objected to, principally from the way in which the capital was to be raised and the parties guaranteed against loss.   The cash was to be raised by a joint stock company, which was to be repaid to the parties by the Colombian Government, in bonds bearing a specific interest from the completion of the work.   This was, in fact, inviting the Government to make another loan for this specific purpose, and, in short, increasing their national debt without appropriate revenues to meet its demands.   One would have thought with a young country that this proposal would have met with immediate sanction ; but on the contrary, the Government, seeing the low state of their finances, and the great difficulties they would have in getting the revenue of the republic to cover the expenditure, trembled at the idea of augmenting their inconveniences, which they even at that time knew must sooner or later plunge the whole country into its present difficulties.   I cannot well explain the unsettled state of the whole of this country, and the fluctuations of opinion which daily take place among the people. One day we hear of nothing but civil war, another brings forward some displeasing decree from Bolivar, whose character as a disinterested man has lost ground very much amongst his own people. The laws in many parts are held in contempt, and a disposition for changing the present constitution is pretty general throughout every department. A division of the republic into states appears inevitable, but the precise basis upon which such a change is to be accomplished is yet undetermined, and probably will remain so for a twelvemonth. If the country had not already suffered severely from internal war, or if the effects were not so fresh in the memory of the present generation, I should say that contention in the shape of war would again break out; but the apathetic disposition of the people, together with the worn-out resources of the nation, will, I think, effectually counteract any such movement.

I was much pleased to learn from your letter of half-a-dozen dates, the arrangements you had made respecting your little daughter, and I hope she enjoys good health, with the whole of your family. I shall be most happy to relate some travellers' stories to her when I return, but I must be careful in my selection, as, if all were told, it might give her ideas a turn too much towards romance.

In the close of your last letter, dated Feb. 2, 1827, you mention that the calisthenic exercises have just come into fashion. This puzzled me not a little. I could not find for the life of me any signification for the new-coined word, and therefore I am as ignorant of the kind of exercise which has become fashionable amongst the ladies as I was before I left England, and I suppose I must remain so until I return.

I was delighted to hear you were studying Spanish, but I am afraid (on my part only) our conversation in that beautiful language must be very limited — 'pero quando nos vemos lo probaremos.'

Quedo su afectuoso amigo,
Robert Stephenson.

Michael Longridge, Esq.
  Bedlington Iron Works, Morpeth, Northumberland.

The Association having notified to him the appointment of his successor, Robert Stephenson, after being entertained at a public dinner, by his coadjutors of all ranks, quitted Santa Ana, and with his friend Charles Empson, who had been his constant associate in his American labours, proceeded to Carthagena to take ship. He had much wished to visit the isthmus before his return to England, but the delay which such a trip would occasion caused him to dismiss all thought of making it. At Carthagena he was joined by Mr. Gerard, an employe of the Association, who was bound for Scotland, having under his charge two little boys, named Monteleagre. Another addition was made to the party in the person of Trevithick, whom Robert Stephenson accidentally met in an hotel. Without funds and without credit, Trevithick, after undergoing indescribable hardships in exploring the isthmus, had made his way foot-sore and almost starved to Carthagena. A strange reverse had come over his fortunes since the time when the Peruvians received him with the honours of a conqueror, and, in anticipation of the fabulous wealth which it was expected would flow to them from his genius, had shod his horses' hoofs with silver. An instructive study was that rude, gaunt, half-starved ' Cornish giant '— eager for fresh knowledge, liberal, daring, self-reliant, and original in all questions pertaining to his own profession, but on all other subjects untaught and unobservant. There is no doubt that the original and daring views of Trevithick with respect to the capabilities of the locomotive made a deep impression on Robert Stephenson.

As there was no suitable vessel about to start without delay from Carthagena for a British port, Robert Stephenson decided to take passage on a ship bound for New York, and thence to proceed to London, or Liverpool. The entire party, including Trevithick, quitted the unwholesome little town of Carthagena, where yellow fever was raging, and set out for New York. The voyage was eventful. At first the weather was serene, and for several days the ship was becalmed amongst the islands. From the stillness of the atmosphere the sailors predicted that on clearing off from there they would learn that a fearful storm had raged in the open ocean. A few degrees farther north, they came upon the survivors of a wreck, who had been for days drifting about in a dismantled hull, without provisions and almost without hope. Two more days' sailing brought them in with a second dismantled hull full of miserable creatures, the relics of another wreck, whom hunger had reduced to cannibalism.*

The voyage was almost at an end, and they had made

As it has been a matter of question whether civilised men in recent times have ever been driven by hunger to cannibalism, the curious and the incredulous will like to have before them Robert Stephenson's account of an occurrence which seafaring men, who *dare* to tell the truth, will admit to be by no means a solitary instance of such horror. 'We had,' Robert wrote from Newcastle on March 1, 1828, to his friend Mr. Illingworth, at Santa Fe de Bogota, ' very little foul weather, and were several days becalmed amongst the islands; which so far was extremely fortunate, for a few degrees farther north the most tremendous gales were blowing; and they appear (from our subsequent information) to have wrecked every vessel exposed to their violence, of which we had two appalling examples as we sailed north. We took on board the wrecks of two crews who were floating about in dis- mantled hulls. The one had been nine days without food of any kind, except the carcasses of two of their companions who had died a day or two previous from fatigue and hunger. The other crew had been driven about for six days, and were not so dejected, but were reduced into such a weak state that they were obliged to be drawn on board our vessel by ropes. A brig bound to Havannah took part of the unfortunate crews, and we took the remainder, having met us near where they were taken up. To attempt any description of my feelings on witnessing such a scene would be useless. You will not be surprised to know that I felt somewhat uneasy when I recollected that I was so far from England, and that we might also be wrecked.' Farther particulars of this tragedy, it may be added, have been obtained from Robert Stephenson's fellow- passengers.

land, when about midnight the vessel struck and instantly
began to fill.   The wind blew a hurricane, and the deck
was crowded with desperate people, to whom death
within gunshot of land appeared more dreadful than
perishing in the open sea.   The masts and rigging were cut
away, but no good was gained by the measure.   Surrounded by broken water, the vessel began to break up,
whilst the sea ran so high that it was impossible to put
off the boats.   By morning, however, the storm lulled,
and with dawn the passengers were got ashore.

Robert Stephenson and his companions naturally pushed
forward in the scramble to get places in the boat which
was the first to leave the sinking ship ; and they had
succeeded in pushing their way to the ladder, when the
mate of the vessel threw them back, and singled out for
the vacant places a knot of humble passengers who stood
just behind them.   The chief of the party was a petty
trader of Carthagena.   He was, moreover, a second-class
passenger, well known to be without those gifts of fortune
which might have made it worth a mate's while to render
him especial service.

On the return of the boat, Robert Stephenson had better
luck, and by 8 o'clock A.M. he was landed, safe and sound,
on the wished-for shore.   Not a life was lost of either
passengers or crew : but when Stephenson and his companions found themselves in New York, they had lost
all their luggage, and almost all their money.   A collection of mineral specimens, on which he had spent
much time and labour, was luckily preserved : but he
lost a complete cabinet of the entomological curiosities
of Colombia, and the box containing his money, on which
his fellow-travellers were dependent.

Fortunately, he found no difficulty in obtaining money in New York. He was therefore in a position to proceed homewards without delay ; but as he was in America he determined to see a little of the country, and to pay a visit to Canada before crossing the Atlantic for Great Britain. At New York Trevithick bade him farewell; but Mr. Gerard, the two Monteleagres, and Mr. Empson, agreed to accompany him on a pedestrian excursion from New York over the border to Montreal.

This arrangement made, Robert Stephenson said farewell to the captain in whose ship he had made the unfortunate passage from Carthagena, and on parting with him asked if he could account for the mate's conduct when the passengers were leaving the vessel. ' I am the more at a loss to find the reason for his treatment of me,' he observed, ' because on the voyage we were very good friends.' ' Well, sir,' answered the captain, ' I can let you into the secret. My mate had no special liking for Mr.·——, indeed, I happen to know he disliked him as strongly as you and the rest of the passengers disliked him. But Mr. —— is a freemason, and so is my mate, and freemasons are bound by their oath to help their brethren in moments of peril, or of distress, before they assist persons not of their fraternity.' The explanation made so impressed Robert Stephenson that he forthwith became a mason, — the master, wardens, and members of the St. Andrew's Lodge No. 7, constituted under the auspices of the Grand Lodge of the State of New York, presenting him (September 21, 1827*) with a document under their seal, in which he is

---

* They had most probably held one or more Lodges of emergency for the purpose of passing him through the several degrees.

styled ' a master-mason of good report, beloved and esteemed among us.'

The master-mason then started for his Northern excursion. A conservative from his cradle, Robert Stephenson,
during his residence in Colombia, had seen the worst side
of republican institutions. The corruption of the Colombian Government was excessive. From high to low, the
bribe and the dagger were regarded as necessary elements of political existence. Of course the venality of
the governing classes and the servility of the mob were
produced by the system that preceded the revolution,
quite as much as by the revolution itself. But however
they may be accounted for, young Stephenson, naturally
averse to liberalism in politics, saw the worst vices of
corrupt despotism openly defended and practised by the
champions of popular opinions. It was natural that he
should leave South America with yet stronger opinions
in favour of vigorous monarchical government. What
he saw in North America did not tend to modify his
views.

On entering New York (he wrote to Mr. Illingworth) we felt
ourselves quite at home. All outward appearances of things
and persons were indicative of English manners and customs ;
but on closer investigation we soon discovered the characteristic
impudence of the people. In many cases it was nothing short
of disgusting. We stayed but a short time in the city, and
pushed into the interior for about 500 miles, and were much
delighted with the face of the country, which in every direction
is populated to a great extent, and affords to an attentive
observer a wonderful example of human industry ; and it is
gratifying to a liberal-minded Englishman to observe how far
the sons of his own country have outstripped the other European
powers which have transatlantic possessions.

We visited the Falls of Niagara, which did not surprise me so
much as the Tequindama. Their magnitude is certainly prodigious; but there is not so much minute beauty about them as
the Salta.

After seeing all that our time would permit in the States
we passed over into Canada, which is far behind the States in
everything. The people want industry and enterprise. Every
Englishman, however partial he may be, is obliged to confess
the disadvantageous contrast. Whether the cause exists in the
people or the system of government I cannot say—perhaps it
rests with both.

The expedition was made on foot, Robert Stephenson
and his companions having with them no apparel save
what they wore and one change of linen. A picture,
painted in 1828, represents the young man as he appeared en route from New York to Montreal, habited in
the variegated poncho which he ordinarily wore in
Colombia, and holding in his hand a straw paramatta hat
with an enormous brim.

One feature of the rural population of the State of
New York greatly delighted him. Their hospitality was
only bounded by their means. Unknown, and apparently poor, wherever the pedestrians halted they were
welcomed to bed and board, and could only rarely induce their entertainers, who usually were little farmers,
or storekeepers, to accept payment for their services.
Often after receiving them for the night, a farmer brought
out his light wagon, and drove them ten or fifteen miles
on their way, and then said good-bye to them, declining
remuneration for his entertainment, his time, and the
wear of his hickory springs.

At Montreal he threw aside his Colombian dress, and,
equipping himself in the ordinary costume of an English

gentleman, went into the best society of the city. After attending a succession of balls and routs given by the colonial dignitaries, he returned to New York, and with his four companions and a servant took his passage to Liverpool in a first-class packet — 'the Pacific.'

At Liverpool he found his father settled in a comfortable house, and superintending the construction of the railway then in progress between that place and Manchester. The years of Robert Stephenson's absence had been years of stern trial to George Stephenson, turning his hair prematurely white, and biting deep lines in his countenance. On September 27, 1825, more than twelve months after Robert's departure for America, the Stockton and Darlington Railway was opened with proper ceremony. The line had been worked with satisfactory results, but still the employment of locomotives on its rails was regarded as little more than an interesting experiment. It was not till the Liverpool and Manchester line was near completion that the real struggle for the use of the locomotive commenced. In the meantime George Stephenson had hard work to maintain his position in the engineering world. The defeat of the Liverpool and Manchester Railway Bill in the June of 1825—a defeat due in a great degree to serious mistakes made by the engineer's assistants in taking the levels for the proposed line — had for a time a most injurious effect on his prospects. Writing to Robert, November 1, 1825, Mr. Longridge observed—

Railways still continue the fashion, though I am sorry to add that your father has not that share of employment which his talents merit. It is expected the Liverpool and Manchester Bill will pass this session ; perhaps an Amended Act will afterwards be procured. The Newcastle and Carlisle Railway Bill will not be brought into parliament until another year. Your father has been employed by the party who oppose this railway, and in examining the line has found greater errors in the levels than were committed by his assistants in the Liverpool Road. Robert! my faith in engineers is wonderfully shaken. I hope when you return to us your accuracy will redeem their character. I feel anxious for your return, and I think that you will find both your father and your friend considerably older than when you left us.

Of the letters which Robert Stephenson received from England whilst he was in Colombia, the majority contained words that caused him lively uneasiness for his father, who was struggling hard to recover ground which had been lost chiefly through the blunders of his subordinates. In 1826 permission was obtained to lay down the Liverpool and Manchester line, and George Stephenson was appointed engineer-in-chief to the undertaking, with a salary of £1,000 a year. It was said by his enemies, and was also thought by some of his friends, that his success in getting the post was only the forerunner of his ruin. Whilst the result of the attempt to make the line across Chat-Moss was a matter of doubt, George Stephenson was generally regarded as being on his trial; and he well knew that in accordance with the success or failure of that attempt, he would be proclaimed a man of stupendous genius, or an ignorant and impudent quack.

With his own profession George Stephenson set himself right sooner than with the public at large. On February 28, 1827, Locke, writing to Robert Stephenson, said—

Since I last wrote you, many circumstances, at that time highly improbable, have occurred; and that shade which was unfortunately cast on the fame of your father has disappeared, and

the place which he must often have reflected on with pain is
now such a scene of operations as sheds lustre on his character,
and will no doubt immortalise his name. All our Directors
are unanimous in placing the utmost confidence in him, which
is certainly the best proof of their good opinion.

Before Robert's arrival in Liverpool at the end of
November in the same year, the shade had indeed
passed from George Stephenson's fame, and the father
and son were able to exchange words of triumphant con-
gratulation as well as of affection.

It was a happy meeting. If the events of the preceding
three years had whitened George Stephenson's locks, and
given him at forty-six years of age the aspect of advanced
life, his head and heart were still young. On the other
hand, his son had changed from a raw Northumbrian
lad into a polished gentleman, having, at an age when
many young men of the upper ranks of English life are
still shirking college lectures and lounging about clubs
and theatres, reaped the advantages of extended travel,
continued mental exertion, and intercourse with men
widely differing in rank, nationality, and experience.

The friend who had shared the perils and trials of
Robert's American life became a guest in George Stephenson's house at Liverpool. When the young men awoke
on the morning after their arrival they found on their
dressing-tables two handsome watches, which had been
placed there whilst they were asleep. In this manner
George Stephenson made good a part of the losses they
had sustained through the shipwreck.

Robert Stephenson had too much business on his hands
to think of making a long stay at Liverpool. With all
speed he went up to London, and had an interview

with the Directors of the Colombian Mining Association, who received him with gratifying expressions of
respect. Though he had ceased to preside over their
interests in South America, they pressed him to continue to give them counsel as to their future operations.
In London he was quickly immersed in business, inspecting machinery, and entering into contracts for the
house of 'Robert Stephenson and Co.' In connection
with a contract and negotiations entered into with a
foreign house he found it necessary to visit Brussels
in December 1827. The journey was purely one of
business; an excursion to Waterloo being almost the
only diversion he permitted himself during the trip.
Christmas Day he spent in London; but with the new
year he was in Newcastle, which for the next five
years was his head-quarters, superintending the factory,
and originating, or developing, those improvements in
the structure of the locomotive which raised it to its
present efficiency from the unsatisfactory position it
held at the opening of the Stockton and Darlington line.

The following letter, written to Mr. Longridge from
Liverpool on New Year's Day 1828, will show how occupied the writer's mind was with the possibility of improving the locomotive.

Liverpool: January 1, 1828.

My Dear Sir,— On my arrival here last Thursday I received
your letter containing the notice of the Darlington meeting on
the 5th instant, which I will attend at your request. I had
hoped that my father would accompany me to the north this
time, but he finds that all his attention must be devoted to this
road * alone.

I have just returned from a ride along the line for seven

---

* i. e. the Liverpool and Manchester Railway.

miles, in which distance I have not been a little surprised to find excavations of such magnitude. Since I came down from London, I have been talking a great deal to my father about endeavouring to reduce the size and ugliness of our travelling-engines, by applying the engine either on the side of the boiler or beneath it entirely, somewhat similarly to Gurney's steam-coach. He has agreed to an alteration which I think will considerably reduce the quantity of machinery as well as the liability to mismanagement. Mr. Jos. Pease writes my father that in their present complicated state they cannot be managed by 'fools,' therefore they must undergo some alteration or amendment. It is very true that the locomotive engine, or any other kind of engine, may be shaken to pieces; but such accidents are in a great measure under the control of engine-men, which are, by the by, not the most manageable class of beings. They perhaps want improvement as much as the engines.

There was nothing new when I left London, except a talk that the Thames Tunnel was about to be abandoned for want of funds, which the subscribers had declined advancing, from the apparent improbability of the future revenue ever being ade-quate to paying a moderate interest. There are three new steam-coaches going on with, all much on the same principle as Gurney's.

Very shortly after my arrival at Newcastle I shall have to set off to Alston Moor to engage some miners, both for the Colombian and the Anglo-Mexican Association.

The New Year therefore opened with an abundance of business for the young engineer.

CHAPTER VIII.

RESIDENCE IN NEWCASTLE.

(ÆTAT. 24–25.)

State of the Locomotive in 1828 — Efforts to improve the Locomotive — The Reports of Messrs. Walker and Rastrick — A Premium of £500 offered by the Directors of the Liverpool and Manchester Railway for the best Locomotive — Mr. Henry Booth's Invention of the Multitubular Boiler—Commencement of the 'Rocket' Steam Engine—A Tunnel across the Mersey—Survey for a Junction Line between the Bolton and Leigh and Liverpool and Manchester Railways — Survey for Branch Line from the Liverpool and Manchester Railway to Warrington—Robert Stephenson's Love Affairs— His Access to Society in Liverpool and London — Miss Fanny Sanderson — Proposal that Robert Stephenson should live at Bedlington—Mr. Richardson's Expostulations—No. 5 Greenfield Place—The Sofa à la mode — Marriage.

THE great and immediate work before Robert Stephenson, when at the opening of 1828 he once more took up his residence in Newcastle-upon-Tyne, was to raise the efficiency of the locomotive so that, on the completion of the Liverpool and Manchester line, it should be adopted by the directors as the motive power of their railway. At that time the prospects of the locomotive were most discouraging. The speed of five or six miles per hour attained on the Killingworth and Darlington lines by no means justified an enthusiastic support of the travelling engines. It was true that they had not been built with a view to speed, but for the purpose of obtaining cheap carriage for coals. Indeed, not many years before, the problem had been to make them move at all. But progression having been accomplished, the next thing was to increase their powers.

No engineer questioned the possibility of improving the locomotive ; but improvement comes slowly, when each experiment leading to it costs several hundreds of pounds. No railway company could be asked to pay for costly trials. That they would use the new machine when inventors and manufacturers had made it a serviceable power was all that could be expected of the directors of railways. As for the public at large, there was amongst all ranks a general opposition to the new method of conveyance. Dislike to novelty, and suspicion of a system not perfectly understood, combined to make enemies for the locomotive. So far was this the case that, notwithstanding the commercial success of the Stockton and Darlington Railway, the Bill for the Newcastle and Carlisle line was obtained in 1829, only on condition that horses, and not locomotives, should be used in working it.

The proprietary of the Liverpool and Manchester Railway shared largely in feelings which were almost universal with the less enlightened multitude. In October 1828, a deputation of the directors visited Darlington and the neighbourhood of Newcastle to inspect the locomotives, and come to a conclusion as to the advisability of employing them between Liverpool and Manchester. ' By this journey,' says Mr. Booth, the treasurer and historian of the Company, ' one step was gained. The deputation was convinced, that for the immense traffic to be anticipated on the Liverpool and

Manchester line, horses were out of question. The debatable ground being thus narrowed, how was the remaining point to be decided? Was a capital of £100,000 to be invested in stationary engines, or were locomotives to be adopted?'

Whilst this question was under discussion, and for several months preceding the October trip just mentioned, Robert Stephenson had been racking his brains to settle another and much more important matter — How to improve the locomotive? how to increase at the same time its power and speed? It was as clear to him, as it had been to his father, that above all things it was requisite to increase in the locomotive the capability of rapidly generating steam. Sufficient heat, with adequate means for rapidly applying that heat to the water, was the desideratum. Eventually the multitubular boiler and the steam-blast of the ' Rocket' gave the required conditions; but previous to their attainment, Robert and his father made numerous failures in attempting to build a really satisfactory travelling engine.

To increase the heating surface, they introduced into the boilers of two engines made for the St. Etienne Railway small tubes that contained water; but the scheme was futile—the tubes soon becoming furred with deposit and burning out. In other engines they with the same object inserted two flues, each with a separate fire. On this principle was constructed ' The Twin Sisters' — the name being suggested by the tubes. A third method adopted was to return the tube through the boiler. A fourth plan — in which may be perceived a nearer approach to the multitubular system — was adopted in a boiler made, at the beginning of 1828, with two small tubes branching off from the main flue. The sketch for this last engine was sent from Liverpool by George Stephenson to his son on January 8, 1828, and in the postscript the sanguine father says— ' The small tubes will not require to be so strong as the other parts of the boiler, and you must take care that you have no thick plates and thin ones, as is often the case with those which come from Bedlington. *You must calculate that this engine will be for all the engineers in the kingdom — nay, indeed, the world — to look at.*'

During his residence at Liverpool, George Stephenson had the great advantage of close personal intimacy with Mr. Booth, the treasurer of the Liverpool and Manchester Railway. Mr. Booth was not only an enthusiastic advocate of the locomotive, but he had a strong natural taste for mechanics, and would probably have distinguished himself had he made engineering a pursuit instead of a pastime. As it is, the multitubular boiler, as a practical agent, must be attributed to him, whatever may be the merit due to such claimants as M. Seguin and Mr. Stevens. Mr. Booth was consulted on all the plans introduced by the Stephensons, and his name continually appears in the letters which passed between the father and son. Writing to Robert, or January 31, 1828, George, referring to *the* experiment then in hand, says—' With respect to the engine for Liverpool, I think the boiler ought not to be longer than eight feet. The engine ought to be made light, as it is intended to run fast. Mr. Booth and myself think two chimneys would be better than one, say eight inches in diameter and not to exceed fifteen feet.' In conclusion the father adds—' I trust the locomotive engine will be pushed.

Its answering is the most important thing to you, and recollect what a number we shall want — I should think thirty.'

On April 15, 1828, George Stephenson, still sanguine as to the result of the boiler with diverging tubes, wrote to Robert —

I am quite aware that the bent tubes are a complicated job to make, but after once in and well done it cannot be any complication in the working of the engine. This bent tube is a child of your own, which you stated to me in a former letter. The interior of a watch looks complicated, but when once well fit up, there needs very little more trouble for one hundred years, and I expect the engine you are fitting up will be something similar to this watch with respect to its working parts.

Five days later George Stephenson, with regard to this same engine, wrote a letter to his son, which is important, as it bears on a question that has been a subject of much warm controversy amongst engineers.

Liverpool: April 20, 1828.

Dear Robert,— I duly received yours dated the 16th inst. I do not think there can be much difficulty in cleaning the refuse matter of the fire from the locomotive-engine boiler. I would make the nozzle pipe that goes in from the blast to be a kind of grating rather than of a conical shape, and to project about two feet into the fire. The grating to be on the upper side. The nozzle piece to be made with a flange, fitting very nicely to the plate at the front of the fire to prevent the escape of air, and kept on by a bolt and cotter, or two screw-bolts. This nozzle piece could easily be taken out at any time and the fire cleaned at the hole. This I think may be done while the engine is working upon an easy part of the road. It appears to me it will be found better to feed one time with coke and the next with coal. I think the one would revive the other.

I do not think there can be so much difficulty in firing on this plan as on the old one.

If you wish me to see the boiler tried before it is put into its seat I will endeavour to come.

If this new engine is found to answer, it will be the bestway to alter all the Darlington engines to the same plan. By doing so the last engine will not be found too heavy for the road.

This engine with the bent tubes, like other attempts made in that year to improve the locomotive, was a failure. Time was running short; the period for opening the new line was fast approaching, and yet George Stephenson and his son had not hit on the way to build such an engine as should sweep the ground from under the advocates of stationary machines.

Writing from Liverpool to Mr. Longridge at the close of the year 1828, Robert communicated the success attending the result of his new boiler made to burn coke.

Liverpool Railway Office: Dec. 1, 1828.

MY DEAR SIR,— It was arranged that I should leave this place to-morrow, but the directors of the Liverpool and Manchester have resolved to-day that my father and I are to meet the deputation which was recently in the north, and enter into detailed calculations relative to the much-contested question of locomotive and stationary engines. Since I wrote you last we have had my new boiler tried at Laird's Boiler Manufactory in Cheshire. You are probably aware that this boiler was made to burn coke. The experiment was completely successful— indeed, exceeded my expectations. Six of the directors went the other day to witness a second experiment. They were all perfectly satisfied. The enemies to the locomotives . . . . said the experiment had answered to the fullest extent. The boilers were shipped to-day in the steam-boat viâ Carlisle, from which place they will be forwarded to Newcastle. . . . . I have had two letters from Forman about the locomotive engine, and he has given us the order at last, but nothing can be done to it until I reach the manufactory.

I am really as anxious to be at Newcastle again as you can be to see me. I cannot say that I like Liverpool. Do not answer

——'s letter until I see you, as he has left me one also, full of such close queries on engineering that I rather hesitate giving him the information in such an offhand manner as he calculates upon.

I am much pleased that you are interesting yourself in the suit of Locomotive *versus* Stationary. It is a subject worthy of your aid and best wishes; but you must bear in mind, wishes alone won't do. Ellis has got settled, and I have got up a proposal in my father's name, which is now before the directors of the Canterbury Railway Co. I expect at a general meeting next Thursday, which will be held at Canterbury, they will decide upon it. I cannot explain it fully in a letter, and therefore defer it till I see you. I have thanked Mr. Booth as you requested.

In January 1829, Mr. James Walker, then of Lime-house, and Mr. James Urpeth Rastrick, then of Stourbridge, were commissioned by the Liverpool and Manchester Railway directors to visit Darlington and the Newcastle country, and report to them on the advantages and disadvantages of the locomotive system. Mr. Walker and Mr. Rastrick were practical engineers of high reputation; and they conscientiously discharged the duties which they undertook. The task assigned them was not to argue on the possibility or probability of speedy improvements in the locomotive. They were to inspect the travelling engines, observe their capabilities, and judge them as they were, not as they might or would be. On the Stockton and Darlington line the two commissioners found locomotives travelling at paces varying between four and six miles an hour. An engine weighing, with its tender, fifteen tons, would drag twenty-three and a half tons' weight of carriages, containing forty-seven and three-quarters tons of goods, at the rate of five miles per hour. So much, and no more, could the locomotive of

1829 accomplish. Of course Messrs. Walker and Rastrick
well knew that the locomotive was in its infancy. Still
they had to concern themselves with the present, and not
the future. On March 9, they delivered in their separate
reports, which recommended the adoption of stationary
engines.*

Robert Stephenson strongly disapproved the reports
He saw in them an obstacle raised to the success of
the locomotive, upon which the extension of the railway
system depended. Writing to a friend on March 11,
two days after the delivery of the hostile reports, he
said — 'The report of Walker and Rastrick has been
received, but it is in favour of fixed engines. We are
preparing for a counter-report in favour of locomotives,
which I believe still will ultimately get the day, but from
present appearances nothing decisive can be said: rely

---

* In his summary of these reports
Mr. Booth says : ' The advantages
and disadvantages of each system,
as far as deduced from their own
immediate observation, were fully
and fairly stated, and, in the opinion
of the engineers themselves, were
pretty equally balanced. The cost
of an establishment of fixed engines
between Liverpool and Manchester,
they were of opinion, would be
something greater than of locomotives to do the same work; but the
annual charge, including interest on
capital, they computed would be
less on a system of fixed engines
than with locomotives. The cost of
moving a ton of goods thirty miles,
that is, from Liverpool to Manchester,
by fixed engines, they estimated at
6·40d., and by locomotives at 8·36d.,
supposing in each case a profitable
traffic *both ways*. But with a system
of locomotives the cost of the first
establishment need only be propor-
tioned to the demands of trade, while
with stationary engines an outlay for
a complete establishment would be
required in the first instance. And
it was further to be considered that
there appeared more ground for expecting improvements in the construction and working of locomotives,
than of stationary engines. On the
whole, however, and looking especially at the computed annual charge
of working the road on the two
systems on a large scale, Messrs.
Walker and Rastrick were of opinion
that fixed engines were preferable,
and accordingly recommended their
adoption to the directors.' — *Henry
Booth's Account.*

upon it, locomotives shall not be cowardly given up. *I will fight for them until the last. They are worthy of a conflict.*'

The 'battle of the locomotive' had indeed begun, and Robert Stephenson was fighting bravely in the contest; but with characteristic prudence he postponed his counter-statement to a triumphant course of counter-action. It was no time for words, at least for words in the shape of a paper controversy. Amongst the directors there was, in spite of the reports, a strong party, if not a majority, in favour of the locomotive. Led by Mr. Booth, and influenced by the enthusiasm of their chief engineer, they gave the most liberal interpretation to the admission of Messrs. Walker and Rastrick, that there was ground ' for expecting improvements in the construction and work of locomotives.' Would it not be well, they asked, to stimulate inventors by a premium to make the expected improvements in time for the opening of the line? The consequence was that on April 20, 1829, the directors offered a premium of £500 for an improved locomotive engine. The following circular announced the conditions and stipulations of the offer:—

Railway Office, Liverpool: April 25, 1829.

STIPULATIONS and CONDITIONS on which the Directors of the Liverpool and Manchester Railway offer a Premium of £500 for the most improved Locomotive Engine.

1st. The said engine must 'effectually consume its own smoke,' according to the provisions of the Railway Act, 7th Geo. IV.

2nd. The engine, if it weighs six tons, must be capable of drawing after it, day by day, on a well-constructed railway, on a level plane, a train of carriages of the gross weight of twenty tons, including the tender and water tank, at the rate of ten miles per hour, with a pressure of steam in the boiler not exceeding 50 lbs. on the square inch.

3rd. There must be two safety valves, one of which must be completely out of the reach or control of the engine-man, and neither of which must be fastened down while the engine is working.

4th. The engine and boiler must be supported on springs, and rest on six wheels; and the height from the ground to the top of the chimney must not exceed fifteen feet.

5th. The weight of the machine, with *its complement of water* in the boiler, must, at most, not exceed six tons; and a machine of less weight will be preferred, if it draw after it a proportionate weight; and if the weight of the engine, &c., do not exceed five tons, then the gross weight to be drawn need not exceed fifteen tons; and in that proportion for machines of still smaller weight — provided that the engine, &c., shall still be on six wheels, unless the weight (as above) be reduced to four tons and a half, or under, in which case the boiler, &c., may be placed upon four wheels. And the Company shall be at liberty to put the boiler, fire-tube, cylinders, &c., to the test of a pressure of water, not exceeding 150 lbs. per square inch, without being answerable for any damage the machine may receive in consequence.

6th. There must be a mercurial gauge affixed to the machine, with index rod, showing the steam pressure above 45 lbs. per square inch, and constructed to blow out a pressure of 60 lbs. per inch.

7th. The engine to be delivered complete for trial at the Liverpool end of the railway not later than the 1st of October next.

8th. The price of the engine which may be accepted, not to exceed £550, delivered on the railway; and any engine not approved to be taken back by the owner.

N.B.—The Railway Company will provide the *engine tender* with a supply of water and fuel for the experiment. The distance within the rails is four feet eight inches and a half.

Never was premium more opportunely offered. It

set engineers throughout the kingdom on the alert.
Now was the time for a house to put itself at the head of
the trade.   If an efficient locomotive could be produced
at the crisis, locomotives would be universally accepted
as the tractive power for iron roads ; and the manufacturers who should produce the engine would, for years to
come, have a monopoly of the best business throughout
Europe.   Robert Stephenson was keenly alive to the
nature of the contest.   Throwing aside his unfinished
criticism of ' the reports ' of Messrs. Walker and Rastrick
till a more convenient time, the young engineer grappled
with the task before him.   As he walked from ' the
works ' to his lodgings, he racked his brains with thinking
what ought to be done.   At times he was despondent.
He had so often felt triumph in the belief that he had discovered how to increase the heating surface of the boiler,
and keep an ever-glowing and fierce furnace in the fire-box.
As often he had been disappointed. The last fifteen months
of his Newcastle labour had been an unbroken series of apparent victories followed by actual defeat. He wrote to his
father ; and for weeks George Stephenson held an ominous
silence.   One morning, however, Robert received a mo-
mentous budget from Liverpool — a design for a new
engine and a letter from his father.   The design was the
original sketch, drawn by Mr. T. L. Gooch, of the multitubular boiler.*   The letter explained the scheme, viz.
to pass heated air, current from the furnace, through nu-

* When Mr. Smiles was engaged
on his biography of George Stephenson, Robert Stephenson gave him
the following account of the origin
of the Multitubular Boiler. The
reader will not fail to remark Robert

Stephenson's characteristic modesty
in passing over, without a word, his
share in the undertaking.
  ' After the opening of the Stockton
and Darlington, and before that of
the Liverpool and Manchester, Rail-

merous small tubes fitted in the boiler and surrounded
with water, and thus, by offering to the water an ex-

---

way, my father directed his attention
to various methods of increasing the
evaporative power of the boiler of
the locomotive engine. Amongst
other attempts he introduced tubes
(as had before been done in other
engines), small tubes containing
water, by which the heating surface
was naturally increased. Two engines with such tubes were constructed for the St. Etienne Railway,
in France, which was in progress of
construction in the year 1828; but
the expedient was not successful; the
tubes became furred with deposit,
and burned out.

'Other engines with boilers of a
variety of construction, were made,
all having in view the increase of
the heating surface, as it then became obvious to my father that the
speed of the engine could not be increased without increasing the evaporative power of the boiler. Increase
of surface was in some cases obtained
by inserting two tubes, each containing a separate fire, into the boiler.
In other cases the same result was
obtained by returning the same tube
through the boiler. But it was not
until he was engaged in making
some experiments, during the progress of the Liverpool and Manchester Railway, in conjunction with
Mr. Henry Booth, the well-known
secretary of the Company, that any
decided movement in this direction
was effected, and that the present
multitubular boiler assumed a practicable shape. It was in conjunction
with Mr. Booth that my father constructed the "Rocket" engine.

'At this stage of the locomotive

engine, we have in the multitubular
boiler the only important principle
of construction introduced in addition to those which my father had
brought to bear at a very early age
(between 1815 and 1821) on the
Killingworth Colliery Railway. In
the "Rocket" engine, the power of
generating steam was prodigiously
increased by the adoption of the
multitubular system. Its efficiency
was further augmented by narrowing
the orifice by which the waste steam
escaped into the chimney; for by
this means the velocity of the air in
the chimney — or, in other words,
the draught of the fire — was increased to an extent that far surpassed the expectations even of those
who had been the authors of the
combination.

'From the date of running the
"Rocket" on the Liverpool and
Manchester Railway, the locomotive
engine has received many minor improvements in detail, and especially
in accuracy of workmanship; but in
no essential particular does the existing locomotive differ from that
which obtained the prize of the celebrated competition at Rainhill.

'In this instance, as in every other
important step in science of art,
various claimants have arisen for the
merit of having suggested the multitubular boiler as a means of obtaining the necessary heating surface.
Whatever may be the value of their
respective claims, the public, useful,
and extensive application of the invention, must certainly date from
the experiments made at Rainhill.
M. Seguin, for whom engines had

tensive heating surface, obtain the means of generating steam with unprecedented rapidity.

At length the problem had been solved. Robert Stephenson immediately was in correspondence with his father as to the details of the new undertaking. It was determined that twenty-four copper tubes should be inserted in the boiler of the new engine, each tube being of a diameter of three inches. In subsequent engines the heating surface was increased with great effect by reducing the size of the tubes, and doubling and even trebling their number. A point, however, was soon reached, where the diminution of the tubes, although it increased the extent of heating surface, had the evil consequence of diminishing the draught from the fire-box to the chimney.*

---

been made by my father some few years previously, states that he patented a similar multitubular boiler in France several years before. A still prior claim is made by Mr. Stevens, of New York, who was all but a rival to Mr. Fulton in the introduction of steam-boats on the American rivers. It is stated that as early as 1807 he used a multi-tubular boiler. These claimants may be all entitled to great and independent merit; but certain it is that the perfect establishment of the success of the multitubular boiler is more immediately due to the suggestion of Mr. Henry Booth, and to my father's practical knowledge in carrying it out.'

  * Unprofessional readers may like to glance at the following lucid explanation of the structure and rationale of the multitubular boiler,

taken from 'Tredgold on the Steam Engine.'

'By causing all the flame and heated air to pass through a great number of small tubes surrounded by the water, a very great and rapid means of heating the water is obtained, as a very large heated surface is thus exposed to the water. The first locomotive engines had merely a large flue passing from the fire-place to the chimney. It was bent round at the end and returned again to the back, the chimney being placed at the same end as the fireplace. The fire was contained in the commencement of the flue, which was made larger for the purpose. This is the general principle of the construction of the boilers for stationary engines, where the size and weight of the boiler are not of so much importance, and the flues can be made

Robert Stephenson was soon busy at work on the new engine, afterwards famous under the name of 'The

---

large enough to get a sufficient area of heated surface in contact with the water. But as in a locomotive engine all the machinery has to be moved at a great velocity, the size and weight of the boiler are obliged to be diminished very much, and some other means has to be adopted to obtain the requisite heating surface.

'The "Rocket" engine, made by Mr. R. Stephenson, which was the engine that gained the prize for the best locomotive at the opening of the Liverpool and Manchester Railway in 1829, was the first engine made with tubes in this country.

'The former locomotives, with only a flue through the boiler, had never been able to travel faster than about eight miles an hour, as they had not sufficient heating surface in the boiler to generate steam for supplying the cylinders more rapidly; the speed attainable by a locomotive being limited only by the quantity of steam that can be generated in a given time. The introduction of tubes into the boiler is one of the greatest improvements that has been made in the construction of locomotives, and was the cause of the superiority of the "Rocket" engine to those that competed with it, and to all former engines. The velocity it attained at the competition trial was 29 miles an hour, and the average 14¼ miles an hour.

'The tubes of the "Rocket" engine were three inches in diameter, and only twenty-four in number. In the engines made subsequently the size was reduced, and the number of them doubled and trebled, by which means the heating surface was very much increased, and with it the power of the engine. The smaller the tubes are, the greater is the heating surface obtained, as small circles have a much larger circumference in proportion to their area than large ones. But when the tubes are diminished in size, the total area of passage through them from the fire-box to the chimney is also diminished; and consequently if the diameter of the tubes were much diminished, the draught of the fire would be checked from the passage to the chimney being too small. The heating power of the boiler would thus be injured, although the amount of heating surface exposed to the water was increased, and the abstraction of the heat from the hot air more perfect. . . . .

'The tubes open into the upper part of the smoke-box, and the hot air passes from them up the chimney. No smoke is produced, except at first lighting the fire, as the fuel used is coke, which does not cause any smoke in burning, but only a light dust. The height of the chimney is obliged to be small, as it can never exceed 14 feet height from the rails; so that the draught produced by it is not at all sufficient to urge the fire to the intense degree of ignition that is necessary to produce steam at the pressure and in the quantity that is required, and some other more powerful means has, therefore, to be adopted to produce the draught. This is done by making the waste steam issue through a

Rocket,' which won the £500 premium offered by the
directors of the Liverpool and Manchester Railway.

The young engineer had, however, other objects of
interest besides the locomotive, in 1828 and 1829.

In the early part of 1828 he was busy constructing machinery for the Colombian Mining Association, and engaging workmen for the mines. In the same year also
he afforded his father personal assistance in superintending
some of the works on the Liverpool and Manchester
Railway. In the March of 1828, he went to Runcorn in
Cheshire to advise on a proposed tunnel under the
Mersey. In June he was at Canterbury. A few weeks
later he was making a survey for the junction line
between the Bolton and Leigh and the Liverpool and
Manchester lines. At the same time, also, he was busy
with a survey for the branch line between the Liverpool
and Manchester Railway and Warrington, eventually the
first line constructed under his sole direction and manage-

---

pipe, called the blast-pipe, which is
directed into the centre of the chimney, and is gradually contracted
throughout its length to make the
steam rush out with more force.
This pipe is made of copper one-
eighth of an inch thick, and is $3\frac{3}{4}$
inches in diameter inside at the bottom where it joins on to the cylin-
ders, and tapers to $2\frac{1}{2}$ inches at the
top.

'The waste steam rushes out of
the pipe with great force up the
chimney, carrying the air with it,
and causing a very powerful draught
through the tubes and the fire. A
whole cylinder full of steam is let
out at each stroke, and the two cylinders deliver their waste steam al-
ternately ; so that when the engine
is running fast, an almost constant
current of steam in the chimney is
produced, and the interval between
the blasts can scarcely be perceived.
By this method the fire is not blown,
as is usual, by forcing air into it, but
by extracting the air from the flues
and drawing air through the fire.
In the first locomotives no means
were used for increasing the draught
of the chimney, and their power of
generating steam was consequently
very limited. The introduction of
the steam-blast for urging the fire,
and of the tubes for conveying the
air through the water, are the principal causes of the great power of
the present locomotives.'— *Tredgold
on the Steam Engine.* Edinburgh,
1838.

ment. These undertakings were ' the trifles ' with which he
filled up the time left on his hands by the superintendence
of the engine manufactory at Newcastle.

Sometimes he fretted under the caprices of directors
and projectors, and once or twice he nearly lost his
temper with a ' board.'

Writing from Liverpool on August 27, 1828, he informed an intimate friend —

I had prepared this morning to get my things packed up for
going off to Newcastle to-morrow morning, but there was a
meeting of the directors of a short line of railway which I have
got the management of near Bolton. The plans and section had
been previously laid before them with an estimate. To-day they
came to a resolution that, although the line pointed out by the
engineer was the best, they were alarmed at the expense of it,
and in consequence ordered a fresh survey and section to be
made, so as to reduce the expense, even at the risk of having a
less advisable line. This is one way of doing things, but proud
as I am I must submit. I have tried in my cool and solitary
moments to look with patience on such proceedings, but, by
heavens, it requires a greater store than I have. I would
patiently bear this alteration if they did it from principle ; but
knowing, and indeed hearing, them say from what the alteration
does really spring, I cannot but consider it unworthy of Liverpool merchants. I plainly perceive a man can only be a man.
As soon as ever he aspires to be anything else he becomes
ridiculous. Come, come, away with moralising thus gloomily.
Affairs go on smoothly in London, at least, the last time I heard
from thence, and as I have not written anything disrespectful
since, they cannot have undergone any material change.

Those who hold that love is merely the amusement of
idleness will find it difficult to account for the fact that
Robert Stephenson at twenty-five years of age, pressed as
he was with various and weighty affairs, found leisure
for indulging the tenderest of human affections. His

father and stepmother had early impressed upon him the advantages of early marriage, and when they endeavoured to withhold him from sailing for Colombia their arguments concluded with an assurance that he ought to be thinking of a wife. In his farewell letter from Liverpool, before starting for South America, he laughingly promised Mrs. Stephenson to marry as soon as he returned to England, after the appointed three years of absence. In America he of course saw but little of ladies' society Beyond an occasional ball at Mariquita he had no means of becoming acquainted with women more cultivated than Señora Manuela, the fat negress who presided over the cuisine of his Santa Ana cottage. His Colombian letters abound with expressions of dissatisfaction at being thus isolated from the poetry and refinement of woman's influence.

On returning to England, he availed himself of every attainable means of entering society. At Liverpool, as well as in town, he was well received in the families of those affluent merchants who were interested in the progress of mechanical science. In many quarters there was a flattering and not unnatural preference shown for him over men his superiors in rank and wealth, by ladies anxious for the establishment of their daughters.

In March 1828, writing to a friend, he said : ' If I may judge from appearances I am to get the Canterbury Railway, which you know is no inconvenient distance from London. How strange! Nay, why say strange, that all my arrangements instinctively regard Broad Street as the pole?'

The attraction in Broad Street was Miss Fanny Sanderson, the daughter of Mr. John Sanderson, a gentleman of good repute in the City. Robert Stephenson had been introduced to the young lady before leaving England for South America, and even at that date he had entertained for her sentiments which, if not those of love, closely resembled them. On returning from Colombia, amongst his first calls made in London he paid a visit to Broad Street, where he met with a cordial reception from Miss Sanderson, and an urgent invitation from her father to be a frequent visitor at his house. He waited some time, however, before he committed himself to the position of a suitor. In the October of 1828 he wrote to a friend, who was also Miss Sanderson's cousin: 'When in London I met my father by pure chance, and as he remained a day I had him introduced to Fanny. He likes her appearance, and thinks she looks intelligent. I took him to the house without her having the most distant idea of his coming. She did not appear confused, and the visit passed off extremely well.'

But it was not till the close of 1828 was near at hand that Robert Stephenson asked the lady to become his wife.

Having made his offer and been accepted, Robert Stephenson did not wish for a long engagement. Indeed, there was no reason for deferring the marriage. Mr. Longridge was very anxious that the young couple should settle at Bedlington ; and Robert Stephenson so nearly complied with his partner's wishes, that he arranged to take a house there, and even made preparations for furnishing it. But to this plan his father and stepmother as well as other friends were so averse that he relinquished the scheme, although the alteration delayed his marriage for some months.

At length a suitable house was found — a small and
unassuming dwelling (No. 5 Greenfield Place, Newcastle).
The surrounding land has, during the last thirty years,
been built upon in every direction, and the inhabitants
of Greenfield Place would at the present date look in
vain from their windows for a picturesque landscape, but
when Robert Stephenson took his young bride there, the
outskirts of Newcastle had few more pleasant places.

Between January and June in 1829, he spent much of
his time in Broad Street.   Wherever he was stationed —
at Liverpool or Canterbury or Newcastle — it was to
London that his thoughts turned, and under the pretext
of ' business ' he made frequent visits to the capital.   The
visits were brief, but they could scarcely be called flying
visits, as the journeying to and fro had to be effected by
stage-coaches.   The men of grave years, given over—
heart, soul, and strength—to business, to whom Robert
Stephenson looked for support, and who had hitherto
regarded him as ' a promising young man,' shook their
heads ominously.   Mr. Richardson, taking a paternal
interest in him, even went so far as to reprove him for
wasting on a pair of bright laughing eyes the time that
might be more profitably spent in paying court to the
magnates of Change.   Robert Stephenson deemed it
prudent to defend himself against the reproaches of the
worthy quaker, who, after reading the exculpatory epistle,
laid it aside to be kept—but not until he had inserted
at the proper place, ' 3 mo. 31, 1829,'—the giddy lover
(in his sane moments most careful to date his letters, and
mark off with a dash the spot on the outer sheet to be
occupied by a seal) having actually omitted to put down
the date.

29 Arundel Street, Strand.

'DEAR SIR,— You do me injustice in supposing that the
ladies in Broad Street engross the whole of my time; I am at
present so ardently engaged in the Carlisle opposition that I
have neither time to visit Broad Street or the Hill (i. e. Stamford
Hill, Mr. Richardson's residence), though a visit to either place
would give me great pleasure. You are really too severe when
you imagine, or rather conclude, that I neglect business for considerations of minor importance. I am well aware that it is
only by close attention to my business that I can get on in the
world. If any appearance of neglect on my part has been
observed by you, I should esteem it a mark of friendship to have
it pointed out by you. The valuation of the mill would have
been forwarded to you immediately on my arrival in London
but for the reason I stated in my last, the 28th. John Dixon
having told me that you thought I was lazy, induced me to
forward it to you in an unfinished state, inasmuch as concerned
the tenor of occupation, which I have not been able to determine
satisfactorily. I saw John Leigh this morning, who it appears
had a lease of the mill from Lord Turner. He mentioned that
some circumstances had removed the lease from his hands, but
on what terms he was holding the establishment was not satisfactorily explained by him. Further than this, I fear I have no
means of furnishing you with the requisite information. There
seems to be some outs and ins which are not easily come at by
ordinary enquiries.

Yours most respectfully,
ROB. STEPHENSON.

As soon as we get through the Carlisle business, I will let
you know when I shall be at Stamford Hill.

In spite of hard work and petty annoyances, however,
he contrived to enjoy himself in London. The preparations for marriage were modest, and precluded all unnecessary expense; for Miss Sanderson had no fortune,
and Robert Stephenson—though he was confident and
hopeful for the future—was far from a rich man. His

principal occupation was the superintendence of a factory which, instead of being a lucrative concern, absorbed
all the money that he and his father could gather together.   So the young people prudently adapted their
expenditure to their means.   They determined to keep
only one domestic servant, and even debated whether
they should spend money on a drawing-room sofa.
Robert Stephenson opposed the outlay as unnecessary,
and therefore bad in principle.   'Reason or no reason,'
he wrote to a friend in Newcastle, 'Fanny will have a
sofa à la mode in the drawing-room.   I shall see you
soon, when we will talk this over.'   Of course the 'talking
over' resulted in his compliance with the lady's wish.
In May the young people shipped from London for Newcastle a piano, which in due course was placed in the
little drawing-room in Greenfield Place.

In June Robert Stephenson went up to London from
Newcastle to be married.   On the 4th of that month,
writing to an old friend, with characteristic frankness he
avowed how profoundly his feelings were moved by the
prospect before him—

I was very much upset (he wrote) when I parted with you on
Gateshead Fell.   So many new feelings and novel reflections
darted across my mind.   It was no ordinary feeling that I was
not to meet you again before my relation, and indeed connection, with the world would be materially changed.   These
sentiments you can appreciate more readily than I can describe
them.

The near approach of his wedding unsettled him for
the performance of business, but did not make him less
anxious to attend to the many calls on his time and care.
The evening before his marriage he received depressing

intelligence of a serious accident to one of the bridges on the Liverpool and Manchester line. On that same evening also he wrote to his good friend, but stern monitor, Mr. Richardson :—

London: June 16, 1829.

Dear Sir,— When speaking of the 'Tourist' steam-packet, I forgot to ask to whom the report of the boilers and flues was to be addressed. I have written to-day full particulars to Dickinson, saying that you would drop a line informing him how to address the report.

I am reluctant to trouble you thus much, but hope you will excuse me. *I am perhaps excusable for neglecting some little particulars last night.* You will have the goodness to inform Mrs. Richardson that, unless something very extraordinary take place, I shall be married to-morrow morning. Afterwards I shall proceed by way of Wales to Liverpool, where I purpose arriving on Monday next.

    I remain, dear Sir,
      Yours most respectfully,
        Rob. Stephenson.

On Wednesday, June 17, 1829, the bells of the parish church of Bishopsgate rang for Robert Stephenson's marriage. As far as bystanders could see, he had made a wise selection of a wife. Mrs. Stephenson was not beautiful, but she had an elegant figure, a delicate and animated countenance, and a pair of singularly expressive dark eyes. A near relation, who knew her intimately from childhood, bears testimony: 'She was an unusually clever woman, and possessed of great tact in influencing others, without letting anyone see her power. To the last her will was law with her husband; but, though she always had her way, she never seemed to care about having it.'

CHAPTER IX.

RESIDENCE IN NEWCASTLE—CONTINUED.

(ÆTAT. 25–28.)

Wedding Trip — Battle of the 'Locomotive' — 'The Oracle' — Construction of the 'Rocket' Steam Engine — The Rainhill Contest — Particulars concerning the 'Rocket' — History of 'the Blast-Pipe' — Triumphant return from Liverpool to Newcastle — Answer to Mr. Walker's Report — Letters to Mr. Richardson — Numerous Engagements — More Locomotives — Opening of the Liverpool and Manchester Railway—Robert Stephenson appointed Engineer to the 'Warrington' and 'Leicester and Swannington' Lines —Discovery of Coal Strata, and Purchase of Snibstone—London and Birmingham Railway—Robert Stephenson employed to carry the Line through Parliament — Opposition to the Line — 'Investigator's' Pamphlet — Robert Stephenson's Evidence before the Lords' Committee — Rejection of the Bill in 1832 — Calumnies — Public Meeting at Thatched House Tavern in support of the London and Birmingham Railway — Bill passes Parliament in 1833 — Robert Stephenson appointed sole Engineer-in-Chief to the London and Birmingham Railway — Leaves Newcastle-on-Tyne — Pupils.

ROBERT STEPHENSON'S wedding trip was a short one. No sooner had he introduced his bride to her new home in Greenfield Place than he devoted all his energies to the superintendence of 'the works,' and especially to the construction of the 'Rocket.' The great and decisive battle of the locomotive, to be fought at Rainhill during the ensuing October, was fast approaching. He had to carry out the instructions which he had received from Mr. Booth and his father. A fearful

responsibility it was for so young a man, still only five
and twenty years of age. He knew that on the result
of the contest his after-success would greatly depend.
The 'Rocket' was to him what ' Chat-Moss ' had just been
to his father. It was a grand trial of his capability as a
practical engineer.

In making the drawings and calculations for the
new engine, he was assisted by Mr. G. H. Phipps, who
recalls with enthusiastic admiration the fine qualities
displayed by his ' chief' at that trying period. Punctual
to a moment, and methodical to nicety, the young
engineer was always at his post, and ready for every
emergency. No mishap found him unprovided with a
remedy. And in laying his plans he did not disdain to
profit by the practical experience of men, who in all that
concerned the science of engineering were mere artizans.
' Come, this is a touchy point,' he would cry good-
naturedly, shaking his head after discussing a difficult
question ; ' let's call in " the oracle." ' ' The oracle ' was
Mr. Hutchinson, a practical engineer, and the superintendent of the factory, to whom the subsequent success
of ' the works ' was greatly due, and who eventually
became a partner in the concern. On his judgement
Robert had such reliance, that he invariably spoke of
him as ' the oracle.' Had Robert Stephenson been an
ordinary man, endowed only with the mere cunning
which often passes current for genius, he would have
picked the brains of ' the oracle ' without letting him be
aware of the operation.

At length the tubes, with their thickened ends brazed
in, were screwed into the ends of the boiler. The work
looked well enough, but no sooner was it tested by

hydraulic pressure than from the extremities of the tubes jets of water flew out upon the dismayed beholders. Here was a conclusion to months of toil and hope. For the first time in the protracted labour Robert Stephenson's self-command gave way, and, hastening to his office, he wrote a hasty report to his father of 'another failure.' Scarcely, however, was the letter posted for Liverpool, when his nobler nature reasserted itself, and he looked about for a way to overcome the difficulty. In a happy moment the right plan occurred to him. The brass screws could not be relied upon, but the copper of which the tubes themselves were made might be trusted. Forthwith he bored, in the ends of the boiler, holes exactly corresponding to the size of the tubes. Into these holes the tubes were inserted, and steel ferrules, or hollow conical wedges, were driven into their ends. By this means the copper of each tube was forced powerfully against the circumference of the hole, and caused to fit perfectly water-tight. The steam having been raised, the result equalled Robert Stephenson's most sanguine expectations, and he despatched another letter to his father, announcing his success. That second letter was crossed on its way to Liverpool by one from his father telling his son to try the very same means which had already proved successful.

The engine was at last taken from 'the works' on Tyne side and conveyed to the Killingworth Railway for trial. Much as there was yet to be effected before the locomotive should be raised to its present state of efficiency, a decided progress had been made. The capability of evaporation had been so raised that, while in the Killingworth engines of 1829 the evaporating power was 16 cubic feet of water per hour, in the 'Rocket' engine, at the

Rainhill experiments, it was 18·24 cubic feet per hour. The vast room still left for improvement may be appreciated, even by an unprofessional reader, when it is stated that the evaporative capability of Stephenson's patent locomotive (of 1849) was 'seventy-seven cubic feet of water per hour, or nearly five times the power of the engine of 1829.'*

After trial at Killingworth, the 'Rocket' was taken to the Tyne and shipped for Liverpool, an insurance of £500 having been effected against the peril of the voyage, which was unusually rough and bad. The vessel arrived at Liverpool so long after her time that she had been given up for lost, and the sum for which the locomotive had been insured had been actually paid to 'Robert Stephenson and Co.' when the ship and her cargo entered Liverpool water safe and sound.

At length October arrived, and on Tuesday, the 6th day of the month, the famous locomotive display at Rainhill began. The story of the competition has been often told, but it is a story that will bear repetition.

The running ground was a dead level, about ten miles from Liverpool, on the Manchester side of the Rainhill Bridge, at a place called Kenrick's cross. The whole country round was alive to the great event. From 10,000 to 15,000 people of both sexes and all ranks assembled to witness the novel contest. To accommodate the ladies, amongst whom was Robert Stephenson's wife — anxious and hopeful for her husband — a booth had been erected at either end of the race-course a few yards from the rails. Bands of music enlivened the entertainment.

---

* Nicholas Wood's 'Address,' 1860.

On the course appeared four locomotive carriages—

No. 1. Messrs. Braithwaite and Erichson's, of London, 'The Novelty,' weight 3 tons 15 cwt.

No. 2. Mr. Hackworth's, of Darlington, 'The Sans Pareil,' weight 4 tons 8 cwt. 2 qrs.

No. 3. Mr. Robert Stephenson's, of Newcastle-upon-Tyne, 'The Rocket,' weight 4 tons 3 cwt.

No. 4. Mr. Brandreth's, of Liverpool, 'The Cyclops.'

Mr. Burstall of Leith had entered his 'Perseverance,' but it did not make its appearance on the 6th, in consequence of an accident which it had sustained on its way to Liverpool.

Mr. Brandreth's ingenious horse-power locomotive was worked by two horses in a frame which, whilst they themselves moved not more than a mile and a quarter per hour, propelled their load of five tons at the rate of five miles an hour. This curious contrivance was an object of general admiration; but as a mere freak of ingenuity, not fulfilling the requisitions of the directors, it of course did not contest for the prize.

The 'Novelty,' 'Sans Pareil,' and 'Perseverance,' not being ready at the appointed time, the race was put off, much to the dissatisfaction of spectators. Two days having been spent in preliminary exercise and mishaps,* 'The first systematic trial of the power of the engines, under the inspection of the judges, took place on Thursday, when Mr. Stephenson's engine, the " Rocket," was brought out to perform the assigned task.' The distance appointed to be run was seventy miles. When fairly started, the engine was to draw, at the rate of at least ten miles per hour, a gross weight of 3 tons for every ton of

---

* 'Liverpool Times,' Monday, Oct. 13, 1829.

its own weight. The prescribed seventy miles were to be accomplished on a level plane of one mile and a half; consequently the course had to be travelled over by the successful locomotive forty times — the same number of stops being made—with consequent loss of momentum which had to be regained.

On Thursday, the 8th, the 'Rocket,' weighing with the water in her boiler 4 tons 5 cwt., began her seventy miles at 10·30 A.M., and accomplished the first thirty-five of them in three hours and twelve minutes. The average rate therefore of this first burst was nearly eleven miles per hour. After a quarter of an hour spent in taking up a fresh supply of water and coke, the engine started again, and accomplished the second thirty-five miles in two hours and fifty-seven minutes, making an average speed of twelve miles per hour. Thus, all stoppages included, the entire time from the commencement to the end of the running was under six hours and a half. At its fullest speed the engine frequently carried its burden at more than eighteen miles per hour, and occasionally it exceeded the rate of twenty miles per hour. It had therefore beaten all previous locomotives, and more than fulfilled the stipulations of the directors.

It remains to speak of the other competing locomotives, the 'Novelty,' the 'Sans Pareil,' and the 'Perseverance.' Scarcely had the 'Novelty' commenced running when an accident to its machinery, or pipes necessitated a stoppage for repair. Another trial, on a subsequent day, was only the occasion of another accident. It was therefore withdrawn from the contest. The 'Sans Pareil,' built by Mr. Hackworth of Darlington, was also unfortunate. On being furnished with its complement of water, it was found to exceed the stipulated weight by 5 cwt. Still, though it was thus disqualified for competition, it was permitted to display its powers over the course. Its speed averaging fourteen miles per hour, with the appointed load, was satisfactory; but an accident stayed its operations at the eighth trip. As for the 'Perseverance,' it was so far inferior to its antagonists— never travelling more than six miles per hour—that its name was scratched from the list shortly after the commencement of the running.

The result was that the 'Rocket' was proclaimed the winner, and the premium was consequently awarded by the directors to Mr. Booth and the Messrs. Stephenson, the former being the inventor of the multitubular boiler, and the latter the manufacturers of the successful locomotive.

One principal feature of the 'Rocket' was the efficiency of its blast, which scarcely in a less degree than the boiler contributed to the victory at Rainhill. With regard to the blast there has been much animated and some acrimonious discussion; and more than one person has been pointed to as the first to devise it. In the first locomotive that ran with smooth wheels on smooth rails —namely, the first of Mr. Hedley's Wylam engines— the waste steam was emitted over the wheels at the side. In the second of Mr. Hedley's Wylam locomotives, built, as the reader recollects, prior to George Stephenson's first locomotive, a different course was employed. To obviate the noise and render the smoke less objectionable, a chamber was constructed in the boiler, into which the waste steam was conveyed from the cylinder by an eduction pipe that was turned upwards. From this chamber the steam in an expanded state passed through another pipe into the chimney. This arrangement precluded anything like an efficient blast, but doubtless the passage of the steam up the chimney, as far as it was in any way influential, quickened the draught. This is a fact which should be remembered. In the second Wylam locomotive the waste steam was emitted into the chimney.

In George Stephenson's first Killingworth locomotive engine the waste steam (either from the first, or at a date shortly subsequent to the completion of the *engine*) was conveyed through a pipe directly into the chimney, without passing through any intermediate receiver; and the noise of the steam forcing its way through the exit pipe and up the chimney, soon procured for the engine the sobriquet of 'Puffing Billy.' No attempt had been made to deaden the noise. There was the blast in unquestionable action, although of trivial efficiency.

In the history of mechanical science there are few points more singular than that the origin of such a power as 'the blast in the steam locomotive' should be involved in obscurity. Amongst the Wylam workmen, it is a matter of firm belief that the ability of the waste steam to quicken the draught through the fire-box was discovered by accident. They state that two workmen, the brothers John and Henry Bell, the one still managing, in the autumn of 1860, a fixed engine at Blaydon, the other driving, at the same date, a locomotive on the Wylam line, effected the discovery in the following manner: — It was their duty periodically to clean the boiler of the Wylam locomotive, and also the exit pipe communicating between the receiver and the chimney. This pipe had a tendency to become furred up, and every time the men scoured off the deposit they also removed some of the metal.   The pipe thus gradually became thin, and in the course of years needed repair.   After the fashion of Northumbrian engine-drivers, who tinker up their engines as unconcernedly as a Suffolk ploughman ties up his horses' tails, the Bells inserted a small rim of iron into the enlarged pipe, thus rendering the mouth far more contracted than it was originally.   The current of vapour passing through the narrow orifice, was, of course, much quickened by the alteration.   Its upward passage was proportionately accelerated ; and with corresponding increase of velocity, the air rushed in from below through the fire-box to fill the vacuum caused by the ascending steam.   So marked was the effect of 'the iron rim' on the speed of the engine, that when the men took their first drive on it, after the work of cleaning and repairing, they were for a few seconds positively alarmed by the speed of their progression.   This is one story.   Another tradition, credited by the present representative of the Stephenson family, is that James Stephenson hit on the secret also by accident.   According to this tradition, James Stephenson, whilst acting as driver, turned the eduction pipe of the first Killingworth engine into the chimney for the purpose of abating the nuisance of the waste steam, which, on being emitted from the side of the locomotive, covered him with moisture and interfered with his line of sight.

Certain it is, that the first Killingworth engine, at a very early date of its existence, had ' the blast ; ' that is to say, the steam went into the chimney in distinct puffs. The assertion that George Stephenson himself 'applied the steam-blast' to his first locomotive *in order* to increase the draught and the heating power of the fire, is improbable. The statement that the blast, when so applied, 'more than doubled the power of the engine,' is unquestionably erroneous — although it was made to Mr. Smiles in all good faith by Robert Stephenson himself. The fact is, the size of the chimney and the small power of the engine, the chimney being altogether out of proportion to the power of a two-horse engine, precluded the possibility of having so efficient a blast. Mr. Nicholas Wood, a scientific engineer, intimately acquainted with the locomotive in question, has publicly stated *—'The blast in the chimney, which afterwards formed so important an element in the evaporation of steam, was then comparatively inoperative, from the imperfect mode in which it was applied, and from the low rate of speed at which the engine moved.'

Of course George Stephenson knew that the tendency of the ascending vapour was to quicken the draught up the chimney. But not the less is it true that the influence of the blast was scarcely appreciable in the Killingworth engines. Years were to elapse before George Stephenson was to awaken to a knowledge of the full capability of the blast. The inability to generate a sufficient supply of steam was, from 1814 to 1829, the reason why the locomotive, instead of being generally adopted on railways, was regarded by sound judges as having only a slight advantage over the stationary engine — an advantage not great enough to secure for it a wide popularity. Throughout the greater part of that time, George Stephenson saw clearly that the two great needs of the locomotive were — more heat, and better means

---

* Mr. Nicholas Wood's 'Address.'

of diffusing that heat. Without a fierce fire, and a large
heating surface, it was impossible to generate the requisite amount of steam. He therefore racked his brains
to invent a boiler offering a wide field of contact for the
heat and the water, and to construct bellows that should
make his fire-box a perpetual furnace. The reader, of
course, bears in mind the agreement between George
Stephenson and Mr. Losh and the Messrs. James as to
boiler tubes, in 1821. In a former part of this work
a letter appears, which shows how George and Robert
Stephenson, in seeking to send an adequate current of air
through the fire of a locomotive, fixed their thoughts on
an artificial and not a natural draught. There are extant
many letters between the father and son, which accord
with the one referred to. Such was the state of things in
1828. Such, too, was the case in 1829, until, *whilst the
'Rocket' was being built*, George Stephenson became alive
to the full importance of a principle which, notwithstanding the structure of his own early locomotives, he had for
fifteen years at least not duly estimated.

During the building of the 'Rocket' Mr. G. H.
Phipps had an engagement at the factory at Newcastle,
having charge of the drawing office, and he was Robert
Stephenson's active coadjutor, and trusted friend. During
a temporary absence of Robert Stephenson from 'the
works,' Mr. Phipps received the following letter from
George Stephenson : —

Liverpool: August 13, 1829.

DEAR PHIPPS,— As I understand Robert is gone to Canterbury, I may mention to you that I have put on to the coke
engine a longer exarsting pipe, riching nearly to the top of the
chimeney, but find it dose not do so well as putting it into the
chimeney lower down. I think it will be best near the level of

the top of the boiler, by doing so it will look neater. the coke engine is doing extremely well—but the ' Lankshire Witch ' is rely doing wonders. A statement of her performance you will see in the paper in a few days.

I am, dear Phipps,
Yours truly,
GEO. STEPHENSON.

Had George Stephenson been for fifteen years aware of the full value of ' the blast ' as a natural bellows, he would scarcely at so late a date have thought of putting the mouth of his ' exarsting pipe ' nearly at the top of the chimney. But it was at this very time — August 1829 — that George Stephenson, whilst he was making experiments on the eduction pipe, to see if the rapid current of its vapour could not be employed with greater effect for the creation of chimney draught, hit upon the full importance of a principle which for years he most probably had regarded lightly.

No time was lost in giving the ' Rocket' the full benefit of the new discovery. When the engine astonished the spectators at Rainhill, the draught of the chimney was accelerated by two blast-pipes. ' Mr. Robert Stephenson's carriage,' says the ' Liverpool Courier,' Wednesday, Oct. 7, 1829, ' attracted the most attention during the early part of the afternoon. It ran, without any weight being attached to it, at the rate of twenty-four miles in the hour, shooting past the spectators with amazing velocity, emitting very little smoke, but dropping red-hot cinders as it proceeded.' The ' Sans Pareil ' had also at Rainhill a very powerful blast, but causes independent of the waste-pipe shut it out from success.

The combination of the multitubular boiler and the blast was most felicitous, and achieved the triumph of the locomotive. They acted and reacted upon each with beautiful effect. A good fire was a necessary condition for the proper action of the multitubular boiler; that good fire was secured by the forcible jets of the exhaust-pipe; those forcible jets were a consequence of the boiler being able to supply the cylinders continuously with steam. Without the blast the multitubular boiler would have been comparatively inoperative; and, apart from the multitubular boiler, a strong, continuous, and regular blast was impossible.

Robert Stephenson went home from Liverpool triumphant. It was a happy meeting between him and his wife in Greenfield Place, whither she had preceded him. He was a made man. Henceforth there was no fear for the locomotive; its speedy and universal adoption had been secured. Not less certain was it that Robert Stephenson and Co. would for many years be the first locomotive manufacturers in the world; but the victory, far from inducing the engineer to relax, only spurred him to increase his exertions. He resolved to lose no time in producing engines superior to the 'Rocket.' Having, however, done so much in the way of professional action, he could afford a little time for professional polemics. As long as the locomotive required him to labour in the workshop, he had abstained from controversy; but now he took pen in hand with the purpose of convincing the public mind that the reports of Messrs. Walker and Rastrick were not supported by the facts which they, previous to the production of the 'Rocket,' had undertaken to examine. It might seem that the time was past for replying to statements which had been exploded by events. But the fact was, Mr. Walker's report had taken a firm hold of the public mind, and its author was by no means disposed to modify his views in deference to recent improvements.

On December 17, 1829, Robert Stephenson wrote to Mr. Richardson —

I am now engaged in preparing an answer to James Walker's report on locomotive and stationary engines. I am induced to do this from the industrious manner with which he has been circulating his report in every quarter of England. He left one with Kingsford, the solicitor at Canterbury, doubtless with some object.

In the February of 1830, therefore, Robert Stephenson, in conjunction with Mr. Joseph Locke, published 'Observations on the Comparative Merits of Locomotive and Fixed Engines.' In this treatise *facts* were closely adhered to, and idle speculation was studiously avoided. Robert Stephenson did not want to startle uninformed readers with the marvels which he hoped to accomplish, but to tell them how much he could· assuredly achieve. He was, therefore, content to say : ' On a level railway, a locomotive engine weighing from four to five tons, will convey twenty tons of goods, exclusive of carriages, at the rate of twelve miles an hour.' The moderation and caution of the writer were characteristics that marked his entire professional career, and contributed in no small measure to his success.

A glance at the following extracts from Robert Stephenson's letters to Mr. Richardson will give the reader a vivid picture of a portion of his professional life during the next few months after the Rainhill contest.

Newcastle-on-Tyne: Dec. 17, 1829.

DEAR SIR,— I was sorry that you passed through Newcastle before I returned from Liverpool, as I had many things to

mention respecting railways which are projected in Cheshire and
Lancashire. . . . The proprietors of the Warrington and Newton
Railways a little time ago proposed a line from the former place
towards Birmingham, but at the outset only intended taking it
up as far as Sandbach, a distance of twenty-two miles from Warrington; the remaining distance to Birmingham is, I believe,
about 53 miles. Should this line go on, it will join the Liverpool
and Manchester sixteen miles from Liverpool, through the medium
of the Warrington and Newton Railway, and will consequently be
of great advantage to both these lines now in progress. I made a
survey about three weeks ago, and lodged the plans in the cus-
tomary manner. This plan or line of communication to Birmingham did not meet the views of the Liverpool people. They
therefore employed Vignoles as engineer to survey a line from
Liverpool to Runcorn, where they proposed making a bridge
over the Mersey at an enormous cost, and in this direction
opening a communication to Birmingham. The Liverpool directors were not agreeable that my father or I should be concerned
in the Sandbach line, as it would be opposed by the Marquis of
Stafford; and as my father might be employed to oppose the line
in Parliament, he and I would thus be brought into direct collision, which would certainly not be very pleasant. Having made
this survey, I was of course bound in honour to sign the plan
and section. What will be the result in Parliament I cannot
guess. There will doubtless be a strong opposition, and perhaps
a fatal one. It is averse to my feelings to be concerned with
any undertaking which might interfere with Mr. Locke's views,
as his kindness to my father has been very great. Being,
however, engineer for the Warrington directors, I could not
refuse with any appearance of consistency to attend to an extension of this line — an extension which, if made, will be of
immense benefit to that which I am now executing. I heard
from Liverpool the other day that another Birmingham line had
been suggested which was likely to obtain supporters. It is to
pass underneath the Mersey opposite Liverpool, continue on
to Chester, and thence to Birmingham, in the same route as the
line my father laid down in 1825. I am not aware of the merits
of this line, but it strikes me that it will be a more expensive
one than that from Warrington by way of Sandbach, and it will
certainly never be of so much importance to the Liverpool and

Manchester line.   There are several other branch railways projected in Lancashire.   The trials at Rainhill of the locomotives
seem to have set people railway mad. . . . . We are getting
rapidly on with four locomotive engines for Liverpool, which I
am confident will exceed the ' Rocket ' in powers.   One of them
will leave here about New Year's Day, and the other three
about the end of January.

Yours very faithfully,
Rob. Stephenson.

Liverpool: Jan. 3, 1830.

My Dear Sir,— On my arrival here I found your letter
written after you left Newcastle.

I wish much I had seen you at Newcastle, were it only for
receiving your instructions concerning the Duke of Norfolk's
coal-field and railway to Sheffield, which I intend visiting on my
way to London the latter end of this month.   I am at present
engaged in getting up the parliamentary plans and estimate for
the Warrington and Sandback railway.   As soon as they are
finished I shall proceed to Canterbury.   By that time I hope
the line will be ready for opening. . . . .

Liverpool: Jan. 25, 1830.

My Dear Sir . . . I have consulted my father on the subject
of the Carlisle end of the railway.   He is quite agreeable to take
the west end of the line and leave it chiefly to my management
for something between £500 and £700 a year.   They would not
expect my whole time to be devoted to it, as an assistant to be
always attending would be requisite; so that it would not
require me to confine my attention to that neighbourhood
entirely.   I should then have the Lancashire and the Warrington
and Newton to attend to.   Amongst them I should divide my
attention, and I see no difficulty in doing that, when I have a
confidential assistant at each place to see that my plans are
carefully and strictly attended to. . . . .

Canterbury: April 28, 1830.

Dear Sir . . . . I regret we are too high for the Darlington
Bridge, but I am afraid we are a great deal too high for the

winding engine at St. Helens, Auckland, but we really cannot
compete with those engine-builders in the neighbourhood of
Newcastle, who not only work for nothing, but who make bad
workmanship. The engine you require for St. Helens is the
same power as one we made for the Liverpool Railway Company,
and will require more workmanship. For the Liverpool engine we
had £1,600, but I daresay you will soon have offers for £1,000;
but it is useless attempting to make engines for such prices,
because I know it is impossible to make a good and substantial
job without reasonable prices.

22 Broad Street Buildings : May 6, 1830.

Dear Sir,— I returned from Canterbury on Tuesday, and
would have answered your letters that day had I not been
unwell.

The Warrington business is closed in the Lords, and the
Leicester committee sits to-day, when my business in London
for this session will be ended. I intend leaving London for
Liverpool, where, according to your letter of the 1st inst., you
will probably be. . . . .

The opening of the Canterbury Railway went off remarkably
well, without a single mishap. The paper will be forwarded to
you by Joshua. I have not seen any detailed acount published.

Still only twenty-six years of age, Robert Stephenson
had made a distinguished position for himself, and every
succeeding year was henceforth to add to his dignity and
worldly prosperity. In the spring of 1830 was opened
the Canterbury and Whitstable line, constructed under
Robert Stephenson's supervision, his father being responsible for the engineering. The same season saw the Bill
for the Warrington Railway safe through both Houses of
Parliament, and Robert Stephenson forthwith began to
construct the line as engineer-in-chief—he having made
the survey, sections, and estimates for the parliamentary
application. In the same session permission was sought

to make another line from Leicester to Swannington ; and
the leave being granted, Robert Stephenson was appointed
principal engineer to that undertaking also.   He had thus
two railways on his hands, whilst at the same time he
continued to direct the operations of the Newcastle
factory, and was actively engaged in improving the locomotive.

The heads of most young men would have been
turned by such a tide of success.   It was, however, remarked that Robert Stephenson did not forget the
modesty of bearing which characterised him in youth.
Indeed, conscious as he was of his power, he in a certain way mistrusted himself, and feared that he might
fail from want of experience, if not from want of innate force.   Whilst he was in London, during the progress of the Warrington Bill through Parliament, he was
accosted by an old comrade of his South American ad-
ventures, whom he had not seen since quitting Colombia.
His friend, of the same age as himself, had recently
returned from America to seek fortune in his native land.
'And here I am back in Old England,' he said, 'looking
about for something to do, whilst the business which fills
your hands is on every man's lips.'   The friends dined
together at an hotel in Bridge Street, and over a bottle
of wine talked of past times, and discussed their future
prospects.   'Of course you congratulate me on my advance towards fortune,' Robert Stephenson said earnestly,
'but I can assure you I sometimes feel very uneasy
about my position.   My courage at times almost fails me ;
and I fear that some fine morning *my reputation may
break under me like an egg-shell !* '

As the works on the Liverpool and Manchester line

were being brought to a conclusion, the directors busied themselves with plans for a public celebration of their labours. In August 1829 Mr. Huskisson visited Liverpool, and was present at an inspection of the line, and at a celebration preliminary to greater rejoicings in the following year. Writing by the hand of his secretary to Mr. Longridge, George Stephenson (August 23, 1829) thus described the preliminary entertainment :—

We had a grand day last Friday — Huskisson visited the greater part of the line with the directors, of course I was one of the party. We first went to the great viaduct, thence along the line to the bridge at Rainhill: then to the commencement of the deep cutting at Olive Mount, where we were met by the locomotive engine, which took the whole party, amounting to about 135, through the deep cutting at the rate of nine miles an hour, to the great delight of the whole party : the engine really did well. We next went to the tunnel, where a train of waggons was in readiness to receive the party. Many of the first families in the county were waiting to witness the procession which, accompanied by a band of music occupying one of the waggons, descended in grand style through the tunnel, which was brilliantly lighted up, the gas-lights being placed at intervals of twenty-five yards. The whole went off most pleasantly, without the slightest accident attending our various movements. Huskisson expressed himself to me highly delighted with what he had seen. Mr. Huskisson and the directors dined with Mr. Lawrence in the evening; the engineer was one of the party, and a most splendid set-out there was, I assure you. The evening was spent in a very pleasant manner.

So pleased was Mr. Huskisson with this demonstration in 1829, that he exerted all his influence to assemble people of high importance to witness the formal opening of the line in the following year. Of that later event the engineer could not say—'The whole went off most pleasantly, without the slightest accident attending our various movements.'

On September 15, 1830, the Liverpool and Manchester Railway was opened with an imposing ceremonial and a disaster that struck to the heart of the country. Amongst those who assembled to witness the event were some of the highest personages of the land, the Duke of Wellington and Sir Robert Peel being conspicuous amongst a crowd of celebrities.

The morning of September 15, 1830, was fine and bright, and the towns of Liverpool and Manchester were in a state of great excitement.   For several days exertions had been made to clear the entire line of obstructions — such as earth-waggons, machinery, and masses of timber — which were collected at various points of the route. The 'points and crossings,' at that time by far the most defective part of the railway system, were all carefully removed, excepting at Huyton (about six miles from Liverpool) and at the two termini, so that with these exceptions there was one unbroken line of rails through the whole way, the risk of carriages leaving the line being thus reduced to a minimum.

At Parkside, the half-way point on the line, adequate preparations were made for renewing the supply of water to the tenders of the engines.   The arrangements for obtaining fresh water not being perfected at Manchester, the requisite supply was provided at Eccles (about four miles distant from the great cotton town) — directions having been given that the engines and tenders should be replenished at that station, after performing the entire journey.   The time occupied by the engines and tenders in running out the four miles from Manchester to Eccles, getting a fresh stock of water, and returning to Manchester, would (it was calculated) be less than the time which the visitors conveyed to Manchester by the trains would require for a lunch provided by the Company in a building adjacent to the terminus. The directors, also, having good reason to fear that persons would put obstructions on the rails, stationed men at intervals along the entire line to see that the way was kept clear.

Every precaution for safety and expedition having thus been taken, the procession was formed of eight trains. The following order of progress was drawn out by Joseph Locke, with the assistance of Mr. T. L. Gooch, his coadjutor in arranging the day's proceedings.

	Directed by	Flagmen	Brakesmen
Northum-brian	George Stephenson	Mark Thompson	Jas. Scott / J. Melling, jun.
Phœnix .	Rob. Stephenson, jun.	Jas. Thompson	James Wood / Hugh Greenshields
North Star	Rob. Stephenson, sen.	W. E. Gillespie	Thomas Harding / Thomas Heaton
Rocket .	Joseph Locke		Jno. Wheatley / Jno. Gray
Dart . .	Thos. L. Gooch .	Saml. Bennet	Jos. Copeland / Jno. Cummins
Comet . .	Wm. Allcard . .	Josh. Richardson	Jas. Cummins / Jno. Melling, sen.
Arrow . .	F. Swanwick .	Jno. Birkinshaw	Gordon M'Leod / Wm. Day
Meteor .	Anthony Harding .	Wm. Gray .	Jno. Harding / Thos. Ilberry

The principal train was drawn by the ' Northumbrian ' engine, under the care of George Stephenson. It consisted of four state-carriages, built for the occasion, open at the sides, and made with the awnings and roofs so high that passengers could walk about with ease. This train, containing the Duke of Wellington, Sir Robert Peel, and other personages of high distinction, was placed alone on the southern line of rails. The seven other trains were placed upon the northern line, an interval of about

six hundred yards being allowed between each train and
the one following it.

The trains were started by bomb of cannon ; and for
the first half of the journey all went well.   At the commencement the speed was slow, but as the carriages passed
through Olive Mount cutting the pace astonished the
thousands who lined the slopes.   Crowds who had assembled at the bridges along the line testified their
satisfaction with renewed cheers.   At Parkside, where a
stop was made for a fresh supply of water, an accident,
however, occurred that altogether changed the character
of the day's proceedings.   Mr. Huskisson, who had made
the journey in the first of the seven trains on the northern
line, left his carriage at the station, and, crossing over to
the state-carriages on the southern line, paid his respects
to the Duke of Wellington, with whom he had for some
time been at variance.   The soldier and the Member
of the House of Commons had just time to exchange
words of reconciliation — the Duke retaining his seat, and
Mr. Huskisson standing on the line — when the ' Rocket '
engine, conveying its train at a moderate pace, swept up,
and bore the latter gentleman to the ground, crushing his
thigh bones.   Without delay the injured man was lifted
into one of the state-carriages, and conveyed at the rate
of thirty-six miles an hour to Eccles, where in the vicar's
house he expired during the evening of the same day.

The dismay of the passengers in the other trains, as
on reaching Parkside they received the sad news, was
followed by uncertainty as to what course it would,
under the circumstances, be best to pursue.   Some
thought it would be more delicate to return to Liverpool
and leave the day's journey uncompleted.   Others, think-

ing of the multitude who awaited their arrival at Manchester, and the panic their non-appearance would create
in that city, argued in favour of proceeding. The debate
lasted so long that an hour and a half slipped away
before the 600 or 700 passengers left Parkside. Finally,
it was decided to go on to Manchester. The engines on
the northern line were once more set in motion — the
three state-carriages on the southern line (one carriage
of the original train together with the ' Northumbrian '
engine was engaged in Mr. Huskisson's service) having
been previously attached by chains to the two leading
locomotives on the northern line. No new difficulty
awaited the expedition until it reached the commencement
of the three miles of cutting, through which the line enters
Manchester. At that point, to the surprise and terror of
the engine-drivers, the slopes of the cutting on either side,
and *the railway itself*, were found in the possession of a
dense mass of people. Through this multitude the trains
had to pass before they could reach Manchester. The
authorities of the town and populous district had taken
the precaution of calling out a large military force to
guard the station from the encroachments of the mob.
But a Lancashire mob is never docile ; and just then
political discontents had made the lower orders especially
unruly. The delay in the arrival of the trains, vague
rumours of a fearful accident, and anxiety to behold ' the
Duke,' whom they cordially detested, had put the excited
populace beyond the control of the military. Pushing
out into the country, the crowds soon outflanked the
soldiers, and took possession of the rails.

It was a trying position for Robert Stephenson, who
headed the procession with the Phœnix engine, to which

were attached the five carriages that constituted its train
at starting.   The state-carriages on the south line (from
the windows of one of which the Duke of Wellington
surveyed the rabble) had been once more annexed to the
'Northumbrian.'   Slackening speed, Robert Stephenson
proceeded cautiously.   But caution had its disadvantages;
for the more reckless of the multitude caught hold of
the carriages and climbed up their sides whilst they were
slowly rolling along.   To complete the confusion, the
political animosities of the mob broke out in acts of
insult and violence.   At various points of the cutting
placards reflecting on the ministry were exhibited, and
weaving machines were set out for inspection with brief
announcements upon them of the prices of labour and
bread.   Brickbats also were aimed at the state-carriages.
Eventually the trains reached Manchester without accident or loss of life; but only to find the station occupied by another mob.   All communication between the
different trains was cut off.   Many of the excursionists
left their seats, and fought their way through the crowd
to the apartments where the Company had provided
lunch.   The Duke of Wellington refused to descend from
his carriage, to which the mob continued to press.   For
some time he kept the rioters in good humour by shaking
hands with their women and children; but the uproar
round the state-carriages increased so much and rapidly,
that, to secure the Duke from risk of assault, it was
thought necessary to remove him from the tumult.
Without much difficulty the 'Northumbrian' engine was
brought round, and the train of state-carriages, containing their due complement of occupants, slowly wormed

its way out of the station and through the excited multitude.

At Eccles, on the return of the Duke's party, another
mishap was added to the tale of accidents. At that
station four of the seven engines from the northern were
found on the southern line, taking in a fresh stock of
water. Had it been possible to carry out the programme,
of course these four engines would have returned to
Manchester and taken up their position on the northern
line before the 'Northumbrian' started; but it had
been found necessary to carry off the Duke without
delay at any cost of trouble and confusion. The consequence was that 'the points and crossings' having
been all removed except at Huyton and the two termini,
there were only two moves on the board open to the
players — either to take the state-carriages back to
Manchester, where the Duke would certainly be received
with insults, and not improbably with a shower of brickbats, or to drive the four engines before the 'Northumbrian' to Huyton (six miles from Liverpool), where they
could be passed over to the northern rails and find their
way to Manchester. The latter course was taken.

The four engines were ordered on before the state-train
to Huyton, and the managers of the remaining three engines of the northern line, after taking in water at Eccles,
conveyed to Manchester the pleasant intelligence that
they were left to convey the excursionists back to Liverpool. It was not till late that these three engines (the
'Rocket,' the 'North Star,' and the 'Arrow') reached
Manchester. Darkness was rapidly coming on. In those
early days of the railway system signal lamps had not
come into use. To lessen the chances of collision, which

were strong in the case of three separate trains following each other closely, the carriages were formed into one train, and the three engines were connected. This long train worked through the crowd, and safely reached Eccles, where the train was stopped for the purpose of enquiring after Mr. Huskisson. On starting again, the couplings of some of the carriages broke, and had to be supplied with strong ropes. At Parkside the train was met by three of the engines which had gone on to Huyton: and these three engines, having at Huyton crossed to the north line, were ready to assist in drawing the carriages. It was determined still to adhere to the plan of having only one train. Two of the recently arrived engines were annexed to the load, and the third engine, the ' Comet,' was directed to precede them at the distance of half a mile, clearing the way before them, and signalling obstacles by holding out a lighted tar-rope.

All went on favourably till Rainhill was reached, when the ascent brought the train to a standstill. To lighten the load the male passengers quitted their seats and walked up the incline, when the engines, relieved of more than half their living burden, managed to get to the summit of the rising ground. On the old racing level of Rainhill a good pace was attained, a wheelbarrow (maliciously placed on the line) being at that point run over and broken to pieces by the pilot-engine. Broad Green embankment and Olive Mount cutting were in like manner passed safely; and the train, after leaving the locomotives at Edge Hill, rattled through the tunnel to the terminus, the mile and a quarter of archway being brilliantly lighted with gas, and the feelings of an anxious multitude, who for hours had been waiting for

the arrival of the excursionists, breaking forth in deafen-
ing cheers.*

An old friend of Mr. Stephenson makes the following
statement :—

'Eventful as Robert Stephenson's life had been, the
year 1830 was perhaps as much marked as any by
important occurrences. In that year the Leicester and
Swannington Railway was commenced, to which undertaking he was appointed engineer, the object of the line
being chiefly to convey to Leicester the coal from the
collieries then existing in the neighbourhood of Swannington.   Early in 1831, during the progress of the
Railway, the Snibston Estate, lying close to the line,
was advertised to be sold by public auction. Forming
his opinion from the geological features of the country,
and from the coal seams which had been already worked
near the surface, on a very limited scale, Robert Stephenson was satisfied that other and more valuable seams
existed under the Snibston Estate.   This opinion he urged
upon his father so forcibly, that the latter persuaded two
of his Liverpool friends, Mr. Joseph Sandars and Sir
Joshua Walmsley, to join him in purchasing the Snibston
Estate. In 1831 operations were commenced, and two
shafts were sunk on the dip, or east side of the estate, but
after getting through the keuper, or new red sandstone
formation, which in that part of the district overlies the
coal measures, they came upon a narrow strip of " green
whinstone," which had descended in a fused state from
the neighbouring volcanic hills of the Charnwood Forest
range. This deposit proved a most formidable obstacle ;

* The foregoing particulars of the
opening of the Liverpool and Manchester Line are taken from a written
communication made by Mr. T. L.
Gooch.

but after a long process of sinking, during which so much time was expended that even Mr. Stephenson's perseverance was nearly exhausted, the sinkers drove a bore-hole through it, and proved the coal measures underneath. This strip of " whinstone " was 20 feet thick, and so hopeless had the task of penetrating it at one time appeared, that a second pair of shafts were commenced to the westward, and in these latter pits this serious difficulty was not encountered. After two years of labour Mr. Stephenson's foresight was rewarded by discovering at a depth of 200 yards from the surface an excellent seam, called the " main coal," which was shortly afterwards worked scarcely more to the advantage of the speculators than to the benefit of Leicester, the inhabitants of which town had in a great measure depended for coal on Derbyshire, the coal being brought to Leicester by canal. Upon the opening of the Leicester and Swannington Railway, the price of coals in Leicester fell nearly 40 per cent., whereby the town gained nearly £40,000 a year. The Snibston Colliery, under the intelligent management of Mr. Vaughan, has proved to be a most lucrative concern.'

In 1830 Robert Stephenson became a member of the Institution of Civil Engineers. In that same year (1830), in consequence of the success of the Stockton and Darlington line, the triumph of the locomotive, and the satisfactory state of works on the Liverpool and Manchester Railway,* a project was revived that had slept for the pre-

---

* A note ought to preserve a story connected with the construction of the Liverpool and Manchester Railway. The excitement of the public mind on the subject of railways gave a new turn to the eccentric humours of disordered intellects. Many were the delusions and extravagances of

ceding five years. As early as 1824 a proposal was made
to lay down a railway between London and Birmingham.
The route of the proposed railway was surveyed in 1825,
but in those hard times for speculative enterprise the
project was set aside. The year 1830, however, witnessed
two proposals, instead of one, for an iron road between
Birmingham and the capital. The one set of projectors
advocated a line by Coventry; the other adventurers
being in favour of a route through Banbury and Oxford.
George Stephenson being applied to for an opinion by
the competing parties, decided in favour of the Coventry
route. The consequence of this decision was that the
rival Companies, instead of aiding the external enemies
who were ready to destroy both of them, prudently joined
their forces, and with united influence applied to Parliament for a line through Coventry. George Stephenson
was at first employed in conjunction with his son as
engineer to make the surveys and plans, and carry the
line through Parliament.

The agreement (signed September 18, 1830) between
the Company and the engineers, stipulated that George

---

persons afflicted with railway mania.
One inoffensive elderly gentleman,
residing in a suburb of Liverpool,
conceived a passion for tunnelling,
and a noble ambition to surpass the
achievements of George Stephenson.
Without making any unnecessary
noise he hired a number of workmen,
and amused himself for awhile with
driving tunnels under the surface of
his own grounds. At length, wish-
ing to astonish the 'professional
hands,' he burrowed beyond the
boundaries of his own property, and
bored into the great railway tunnel,
then near its completion. George
Stephenson had heard nothing of
the monomaniac's proceedings; he
was, therefore, not a little astonished
one morning, as he passed along his
tunnel, to hear a noise of a human
voice cheering over his head, and on
looking up to see, through a hole
knocked in the brickwork of the
tunnel, the protruded face, glowing
with exultation, of an elderly gentleman, who persisted in demon-
strations of satisfaction and excla-
mations of 'How are you?'

Stephenson was to receive for his time actually expended on the work seven guineas per diem, and Robert Stephenson five guineas, free of all expenses. George Stephenson's appointment, however, was little more than nominal. The surveys were made by Robert Stephenson, who in the subsequent parliamentary battles was the engineering authority of the projectors, and ultimately, on the Bill being obtained, was made ' engineer-in-chief ' for carrying out the works, his father being in no way whatever associated with him. It is right that this fact should be borne in mind, as a succession of writers have credited George Stephenson with the construction of the first of our existing 'great railways,'—the first railway connecting London with a distant seat of industry. In some inaccurate works the ' London and Birmingham Railway ' is spoken of as having been constructed by George Stephenson, in others by George Stephenson and Son, in others by Messrs. Stephenson. The line was, however, constructed by Robert Stephenson alone, and to him is

---

* As public interest may be gratified, and undesirable discussion obviated, by the publication of the agreement just alluded to, it is here printed.

'Birmingham: Sept. 18, 1830.
' Memorandum of Agreement entered into between Messrs. George Stephenson and Son, of the one part, and the Committee of the London and Birmingham Railway Company, of the other part.

'First, the said George Stephenson and Son undertake and agree, so far as their best and utmost exertions will enable them, to make the ne-

cessary plans, sections, and book of reference for the proposed railway from Birmingham to London, and to do everything that is necessary for that purpose in time to comply with the Standing Orders of the House of Commons, so as to enable the solicitors to insert the necessary notices in the newspapers during three weeks before the sitting of Parliament, and to affix necessary notices on the doors of the several sessions houses at the next Quarter Sessions, and to deposit the plan and book of reference, &c., with the clerks of the peace of the several counties, and in the Private Bill Office, on or before the 24th day of October next, and in due the entire merit of overcoming all the gigantic obstacles to its construction.

Robert Stephenson made three distinct surveys for the London and Birmingham line, besides several minor surveys of different portions of the country, for the purpose of ascertaining whether the route could not be improved. The first survey was made in the autumn of 1830. In 1831 a second line was marked out, almost identical with the one eventually executed. The plans and sections having been deposited, and the requisite amount of shares subscribed for, an application was made to Parliament, and a Bill to enable the Company to make their proposed railway was read the first time on February 20,

---

every other respect to comply with the Standing Orders of the Houses of Parliament, so far as the duty of an engineer extends.

'In consideration of which the Committee agree to pay to Mr. George Stephenson the sum of seven guineas a day during the time that he shall be occupied in the business, and to Mr. Robert Stephenson the sum of five guineas a day during such time as he shall be employed in the business, and to pay the usual charges to surveyors employed by Messrs. Stephenson and Son, and to pay to Messrs. Stephenson and Son the usual travelling expenses.

'And the said George Stephenson and Son agree that Mr. Brunton shall be the resident engineer at the London end, and fully undertake and bind themselves not to be concerned in any line of railway whatever that can be injurious to this Company's line, or any part of it, during such time as they are employed as engineers to this Company.

'(Signed) John Corrie, on behalf of the London and Birmingham Company.
Geo. Stephenson.
Rob. Stephenson.
'Witness to the signature,
'Josiah Corrie.'

The appointment made legally binding by this unartistic and loosely drawn agreement, was an appointment of George and Robert Stephenson to lay out—*not* to construct and make—the contemplated line. It referred only to the survey and parliamentary engineering. In the following year it was superseded by another agreement. It was, therefore, in reality a most unimportant feature of the history of the London and Birmingham line; but it has misled numerous writers into thinking that the elder Stephenson was united with the younger in designing and carrying to triumphant completion the vast engineering operations on the railway in question.

1832. The third survey was made in the autumn following the last date.

The two first surveys were no slight addition to the labours and responsibilities of a young engineer, with the construction of two lines of railway already on his hands, besides the superintendence of a large engine-factory in Northumberland, and extensive mining operations in Leicestershire. In every parish through which Robert Stephenson passed, he was eyed with suspicion by the inhabitants, and not seldom was menaced with violence. The landed gentry were not alone in expressing aversion to a set of men tramping through their fields, and proposing to drive a road, with their leave or without it, across their property. The aristocracy regarded the irruption as an interference with territorial rights. The humbler classes were not less exasperated, as they feared the railway movement would injure those industrial interests by which they lived. To the residents of a market town on a turnpike road, with its ten or fifteen coaches per diem, dropping passengers at its chief hotels, a railway engineer and a ruiner of trade were convertible terms. 'Suppose railroads answered,' asked critics, 'what would be the result? would not the wealthier residents of the neighbourhood invariably travel up to London to make their purchases, and leave the poor country shopkeepers to starve?'

Nor was the opposition confined to the rural population. In London, journalists and pamphleteers, whilst they professed to discuss the new project dispassionately and 'without prejudice,' distributed criticisms which at the time of their delivery were manifestly absurd, and prophecies which time has signally falsified.

'Investigator'* (in 1831), taking for the motto of his pamphlet 'No argument like matter of fact is,' undertook to prove by *facts*, that a railroad between Birmingham and London could not answer. The success of the Liverpool and Manchester Railway he accounted for by the peculiarities of the trade between those two towns, and maintained that the same system of locomotion which was admirably adapted for bales of cotton wool would fail when employed for the general convenience of the public.

Touching on the dangers and inconveniences of railway travelling 'Investigator' says —

That there are other dangers, and most formidable ones too, besides accidents to the engine, there have been too many and too melancholy proofs on the Liverpool and Manchester line. There was the late Mr. Huskisson, there was the engineer's own brother, and there has been a number of others; the amount of whom there is said to be considerable reluctance in disclosing. In short, during the few months that elapsed between September 15, and December 15, 1830, there occurred more fatal accidents upon the thirty-one miles of railway between Liverpool and Manchester, than upon all the road between Birmingham and London in as many years.†

The causes of greater danger on the railway are several. A velocity of fifteen miles an hour is in itself a great source of danger, as the smallest obstacle might produce the most serious consequences. If, at that rate, the engine, or any forward part of the train, should suddenly stop, the whole would be cracked by the collision like nutshells. At all turnings there is a danger

---

* Remarks on the Proposed Railway between Birmingham and London, proving by Facts and Arguments that that Work would cost Seven Millions and a Half; that it would be a Burden upon the Trade of the Country, and would never pay.

'No argument like matter of fact is.' By Investigator. London: T. M. Richardson, 23 Cornhill; J. Ridgway, 169 Piccadilly.

† The reader must bear in mind 'Investigator's' motto —'No argument like matter of fact is.'

that the latter part of the train may swing off the rails; and, if that takes place, the most serious consequences must ensue before the whole train can be stopped. The line, too, upon which the train must be steered admits of little lateral deviation, while a stage coach has a choice of the whole roadway. Independently of the velocity, which in coaches is the chief source of danger, there are many perils on the railway: the rails stand up like so many thick knives, and anyone alighting on them would have but a slight chance for his life. On a road crowded with engines, the escape from the rails would avail him but little, as before he could recover himself from a slight stunning, a train on the next rails would be up, and before the conductor could arrest the progress of that he would be cut asunder. . . . .
. . . Another consideration which would deter travellers, more especially invalids, ladies, and children, from making use of the railways, would be the want of accommodation along the line, unless the directors of the railway chose to build inns, as commodious as those on the present line of road. But those inns the directors would have in part to support also, because they would be out of the way of any business except that arising from the railway, and that would be so trifling and so accidental that the landlords could not afford to keep either a cellar or a larder.

Commercial travellers, who stop and do business in all the towns, and by so doing render commerce much cheaper than it otherwise would be, and who give that constant support to the houses of entertainment which makes them able to supply the occasional traveller well and at a cheap rate, would, as a matter of course, never by any chance go by the railroad; and the occasional traveller, who went the same route for pleasure, would go by the coach-road also, because of the cheerful company and comfortable dinner.

Not one of the nobility, the gentry, or those who travel in their own carriages, would, by any chance, go by the railway. A nobleman would really not like to be drawn at the tail of a train of waggons, in which some hundreds of bars of iron were jingling with a noise that would drown all the bells of the district, and in the momentary apprehension of having his vehicle broke to pieces and himself killed or crippled by the collision of those thirty-ton masses.

An unfair attempt has been made in various quarters
to throw obloquy on the aristocratic classes of the
country, by representing them as the especial opponents
of the earlier railways. As the chief owners of property
required by the projectors of the new roads, the functions
of opposition were principally discharged by them; but
their antagonism to the novel system was admired and
encouraged by all sections of society. Corporations of
provincial boroughs, tradesmen of petty towns, small
yeomen, trustees and mortgagees of turnpike tolls, in
short, all holders of vested interests, were zealous to
crush at their first appearance undertakings which were
sure to disturb and not unlikely to prejudice existing
arrangements. Small proprietors fought against the
Stephensons to their utmost. The great ones of the earth
could do no more. At this date the reader laughs at
' Investigator's' arguments and fears; but thirty years
since, before railways were affairs of familiar knowledge,
many a reader who now despises ' Investigator' would
have thought him very clever, sound, and practical.

In spite of the prevailing antagonism to railways the
Bill for the London and Birmingham line passed the
Commons in 1832 after hard fighting. In the House of
Lords, however, the result was different. The Lords'
Committee came to the conclusion, 'That the case for
the promoters of the Bill having been concluded, it does
not appear to the Committee that they have made out
such a case as would warrant the forcing of the proposed
railway through the lands and property of so great a
proportion of the dissentient landowners and proprietors.'

In the parliamentary progress of the Bill, Robert
Stephenson was subjected to searching cross-examination,

but, ready as well as resolute, quick as well as patient, he was a difficult witness for opposing lawyers to deal with.*

---

* The following extracts from the evidence will give the reader some idea of the duels fought on this occasion between Robert Stephenson and Mr. Harrison, who, in conjunction with Mr. D. Pollock, appeared as counsel for the trustees of the tolls of the Sparrow's Herne Road.

' *Cross-examined by Mr. Harrison.*

' " That piece of board would also have to be put on the top of the embankments ? "

' " It would."

' " The line given in the section deposited with the clerk of the peace, is not the line that will ultimately be the line of the railroad ? "

' " It is the surface of the embankment before the metaling is laid on."

' " Any embankment would be two feet higher than what is described ? "

' " Yes."

' " Have not marks been put in describing the line of the surface ? "

' " Yes."

' " And have they not since been scratched out ? "

' " I believe not."

' " Were they on the section deposited in the other House ? "

' " Yes."

' " Any individual who looks at the section will see that this line is ultimately not to be the surface line by two feet ? "

' " It is the line that is always put on the section to represent the surface invariably."

' " Do you mean to state that ? "

' " I do not know a single exception to it."

' " Do you mean to state that the line so marked on the section has not always invariably been considered at all times as the surface line ? "

' " Certainly, in my practice it never has."

' " Then how came you to describe it as the surface ? "

' " It is the surface of the embankment before the rails are put on it."

' " Will not common people suppose that the surface of the embankment means the top of the place on which the things travel ? "

' " I do not think any engineer would ever so consider it; they always look on the line laid down in the sections as top of the cuttings."

' " If every one of the engineers have made a mistake in supposing the direct contrary, should you attribute it to their ignorance ? "

' " There was not a single discussion that took place on that point with me. The engineers that were employed to make the estimate went through the estimate without asking a question as to what the line represented."

' " That is not an answer to my question. If all the engineers employed by the opposers of the Bill to examine the line as to the injury done to their land, and estimate the quantity of it taken by the embankments —if they have all considered that that is the surface on which the carriages run, do you attribute that to their ignorance ? "

' " Not to their ignorance, but to a disposition to increase the cost of the work."

It was, however, a trying ordeal — trying alike to his temper and his knowledge. His want of professional

---

' " Do you think that they would wilfully do that ? "

' " I do not believe that any impartial engineer would suppose the line to mean anything different to what I suppose it to mean." '

On a subsequent day Mr. Harrison, having in vain endeavoured to show that Robert Stephenson's 'borings' were wrong, and his estimates for bridges and arches altogether incorrect, proceeded to throw out insinuations which were immediately seized upon by Dr. Lardner, and formed an important feature of his well-known attack upon the Stephensons in the 'Edinburgh Review' for October, 1832.

' " In page 115 there is a clause preventing any horse being driven or ridden on the railroad. Is that to preclude the use of horses in dragging carriages on the railroad ? "

' " I suppose it is."

' " Are persons to be allowed to draw railway carriages by horses ? "

' " I conceive that is totally incompatible with a line where locomotive engines are used."

' " All the engines are to be approved by the Company ? "

' " Yes."

' " The engines are not to be exclusively furnished by the Company ? "

' " No."

' " Are you and your father great manufacturers of steam-engines ? "

' " We do manufacture them."

' " For this railroad, do you not, and for others ? "

' " For any railroad."

' " You are the persons who are to be applied to ? "

' " I do not know that that may be the case. *I am only engineer for the time being."*

' " As long as you and your father are the engineers, you are the people to decide what engines shall be used by other people ? "

' " If there is any fear of prejudiced opinions being given by us, I apprehend the committee may easily find a remedy by calling in other engineers to decide the point."

' " There would be no remedy in the Bill ? "

' " He would still be the Company's engineer."

' " The engines to be used are to be approved of by the engineer for the railroad company; and there is no appeal from this decision, if he says he does not like the engines ? "

' " I may not be the engineer for the Company. I apprehend I may be done with the railroad as soon as the railroad is completed, if I am there so long. I may not be the resident engineer." '

On the next day Mr. Harrison resumed his cross-examination of the young engineer, whom, in his zeal for his clients, he had depicted as one who would recommend his own inferior locomotives, and exclude by base influence the superior productions of rival manufacturers. On the present occasion the counsel endeavoured to break down Robert Stephenson's evidence on the subject of 'friction,' by the statements advanced in his 'Answer' to Mr. Walker's report three years before.

' " You stated from sixteen to twenty feet of elevation is equal to a mile. Is it not more than that ? "

experience was superciliously suggested, his answer was
a curt statement of *what he had done.*   It was insinuated

' " I am quite convinced it is not
more."

' " Then it is not twenty-six ? "

' " I am quite sure it is not."

' " I only caution you, in order
that you may be supposed to commit
yourself too hastily to that answer.
Do you stand by that answer ? "

' " That elevation — that is, equal
to a mile going round.   It depends
very much on the friction of the
wagons employed on the railroad."

' " What is the friction on a
level ? "

' " It varies from 6 lbs. to 9 lbs.
What I consider 8 lbs. is amply sufficient for friction on level ground.
I am quite sure I am overstating the
thing when I say 8 lbs."

' " Eight is equal to the friction
for a ton, you say ? "

' " Eight is supposed to be so.
You might overhang a pulley and it
would draw a ton.   The experiments
I have made with wagons make it
considerably less.   They are liable to
get out of order."

' " You abide by your answer,
and state distinctly it is not twenty-
six feet, but sixteen and twenty ? "

' " Yes, I do."

' " Now you published some calculations on it, did you not ? "

' " No, I did not."

' " Did you not publish some experiments in reply to the report made
by Mr. James Walker ? "

' " Yes."

' " Was it not put at 1 to 200 on
a plain surface ? "

' " If I had taken the greatest
advantages of the circumstances that
were allowed me in reply to Mr.
James Walker's report, I should

have stated it at eight.   He took it
at twelve or thirteen : upwards of
thirty per cent. more than experiments proved to me to be correct."

' " He took it at twelve.   Did you
not in your answer put it at 1 in
200 ? "

' " Yes; which is the friction
actually existing on the coal wagons
in the north of England on a very
bad railroad, when compared with
good ones.   On the Liverpool and
Manchester I am quite sure there is
not a wagon moving with the friction
of 8 lbs. to a ton."

' " What would that be instead of
1 to 200 ? "

' " It will be 1 to 280."

' " Instead of its being 1 to 200,
the calculation you put it at, you
now put it at 1 to 280 ? "

' " It is calculated to give an unfavourable impression.   The reason
I put it at 1 to 200 at that part was
this — that the wagons on the railroads in the north of England are
employed there with axle-trees of
considerably larger size than in the
Liverpool and Manchester.   There
the bearings are put on the outside
of the wheels; the size of the axle-
trees are reduced to each one inch
and three-quarters.   For coal wagons
in the north of England the diameter
of the axle-trees is a full three
inches."

' " These are reasons for taking it
at 1 to 280 ? "

' " Yes; from the improvement
in the wagons."

' " The construction of wagons in
the north creates a greater degree of
friction ? "

' " Most decidedly."

that he and his father would supply the petitioning
Company with inferior locomotives, and shut out from
competition the superior engines of rival manufacturers.
His reply was that the Company would know how to
take care of themselves.

Of the exertions made by Robert Stephenson to get
the Bill through Parliament the following story will afford

———————

' " That was the report of Mr.
James Walker on the Manchester
and Liverpool Railway ? "

' " Yes, it was."

' " The Court of Directors desired
him to take into consideration the
difference between locomotive power
and stationary engines ? "

' " Yes."

' " He made a report, on which
you made observations, putting it at
1 to 200 ? "

' " Yes."

' " You would not wish to be
bound by that ? "

' " *Certainly not; I am making
calculations on recent facts.*"

' " At that time you took from
the Liverpool and Manchester ? "

' " Mr. James Walker went to
the north of England by way of
examining circumstances connected
with locomotive engines travelling
and stationary engines working
there. He took the friction at twelve
and a half; and he cannot produce a
single instance where it was twelve
and a half. I took the friction of
wagons in the district he had been
examining." '

Unable to shake the witness, the
counsel proceeded to suggest that so
young a man necessarily lacked experience, and was, indeed, merely
his father's puppet.

' " Have you ever," was the enquiry, " constructed a public work
of that sort yourself ? "

' " Yes; the Warrington Railway
and the Leicester Railway."

' " What length is the Warrington ? "

' " Only five miles."

' " Is it constructing or constructed ? "

' " It is completed some time."

' " What is the length of the
Leicester ? "

' " Sixteen miles."

' " Is it now at work ? "

' " More than twelve or thirteen
miles of it ought to have been
opened yesterday, but they deferred
it till the 9th of this month."

' " Between what points is it ? "

' " Between the town of Leicester
and the coal field of Swannington."

' " Where is Swannington ? "

' " Near Ashby de la Zouch.'

' " The only two you constructed
yourself ? "

' " The only two under my own
charge." '

(*Vide Minutes of Evidence taken
before the Lords' Committees to whom
was referred the Bill, intituled, 'An
Act for making a Railway from
London to Birmingham.'*)

an example.   The opposing counsel directed all their
powers before the Commons' Committee to show that
Robert Stephenson was ignorant of the geological conditions of the country, and consequently proposed to
make his cuttings through the Tring Ridge at so small an
angle that the sides of the excavation would fall in upon
the way.   The argument of course was that, since this
mistake had been made by the engineer, the estimates
were enormously beneath the sum required for the
undertaking, as the increase of the angle of a cutting
would greatly increase the labour and expense at which
it could be completed.   It was to no purpose that Robert
Stephenson offered to stake his reputation that his calculations were reliable.   The barristers ridiculed his as-
surances, and the Committee were evidently impressed
by the objection.   Leaving the Committee- room with his
examination still unfinished, though he had been subjected
for three days to a cross-fire of questions, Robert Stephenson took counsel within himself what he should do.   He
remembered that there was at Dunstable a cutting through
the same formation.   The cutting was Telford's work.
How could he ascertain the angle of Telford's cutting ?
How could he establish the point ?   The question was
soon answered.   He had not been in bed for four nights,
and he had work before him that would keep him in
town till past midnight ; but nevertheless he determined
to visit Dunstable before again entering the Committee-
room.   At midnight he supped, and then had a short
nap, from which he roused himself to get into a post-
chaise with his friend Mr. Thomas Longridge Gooch.
By dawn the two young men were at Dunstable.   By
ten o'clock they were in counsel's chambers in London,

with the intelligence that they could go into the Committee-room and testify that the angle of Telford's cutting,
which had stood the test of time, was the same as the
angle of the cuttings provided for in the estimates.

But toil, patience, forbearance, were all thrown away.
The result of the enquiry, foreseen as it was by those
who were better acquainted with the animus of the
Committee, had not been anticipated by Robert Stephenson, and he was deeply chagrined at the rejection. His
mortification was so manifest that Lord Wharncliffe, the
chairman of the Committee, took him aside and said with
characteristic kindness, ' My young friend, don't take
this to heart. The decision is against you ; but you
have made such a display of power that *your fortune is
made for life.*' These words of sympathy and commendation, coming from a nobleman who, as one of the
' grand allies,' had been amongst his father's earliest
employers and patrons, went to the young man's heart,
and with emotion he often recalled them in after life,
when he reviewed the earlier battles of his career, or
himself held out an aiding hand to struggling merit.

The adverse decision which called forth Lord Wharncliffe's generous sympathy was the signal for the enemies of
the two Stephensons to renew their efforts to make both
father and son the objects of public suspicion. Robert
Stephenson was no exception to the rule that envy is the
shadow of success. At this date it would be an un-
grateful and a useless task to drag into notoriety the
persons who from private pique or professional jea-
lousy used unworthy means to lower the reputation of the
two greatest engineers of this or any other age. Robert
Stephenson wisely paid no attention to malicious rumours.

But when a distinguished scientific writer, who had been misled by detractors, availed himself of his position, on the staff of the 'Edinburgh Review,' to give the stamp of authority to erroneous statements, Mr. Charles Lawrence, the chairman of the 'Liverpool and Manchester Railway,' officially published a complete refutation of the writer's groundless accusations.*

Notwithstanding the rejection of their Bill, the projectors of the London and Birmingham Railway Company were not disheartened. On Friday, July 13, 1832, the first Friday after the rejection of the petition, a public meeting of persons favourably disposed to the Railway was held at the Thatched House Tavern. Sixteen peers and thirty-three members of the House of Commons were present. The chairman of the Commons' Committee was one of the representatives of the Lower House, and Lord Wharncliffe, the chairman of the Lords' Committee, presided at the meeting. Two resolutions were put and carried unanimously. The first resolution, moved by the Earl of Denbigh, and seconded by Sir J. Skipwith, M.P., was —

That, in the opinion of this meeting a railway from London to Birmingham will be productive of very great national benefit.

The second resolution, moved by the Earl of Aylesford, and seconded by Sir Edward D. Scott, Bart., M.P., was—

That the Bill for effecting this important object having passed the House of Commons after a long and rigorous examination of its merits, it must be presumed that its failure in the

---

* Liverpool and Manchester Railway. Answer of the Directors to an article in the 'Edinburgh Review' for October 1832. Liverpool, 1832.

House of Lords has arisen from apprehensions on the part of landowners and proprietors respecting its probable effect on their estates, which this meeting firmly and conscientiously believes to be ill-founded.

This demonstration had an immediate effect on the country. It was felt by those who had opposed the measure from jealous anxiety for the interests of property, that they had not much to fear from the new road, when landed proprietors of high character and hereditary possessions could be found to support such resolutions. It was learnt also that Lord Sefton and Lord Derby, the strenuous opponents of the Liverpool and Manchester line, had become so far converts to the railway system as to have been supporters of the London and Birmingham project. The opposition, which refused to be influenced by such authority, was found not unwilling to listen to other considerations. The bribe reached where reason could find no entrance. In some cases enormous sums of money were paid for the acres of obstinate landowners. The consequence was that in the session of 1833, on the renewal of the petition (Robert Stephenson having in the meantime made a third survey of the line) a Bill was obtained, giving the directors power to construct their line.

It now remained for the directors to appoint an engineer for the accomplishment of the task. Robert Stephenson had high hopes of getting the post. His energy in making the survey, and his conduct as a witness before committees, had won for him many new and powerful friends. But he was still young—*very* young—to be engineer-in-chief to such an undertaking. In the directory, there were of course several persons who honestly mistrusted *young* genius.

Writing to Robert Stephenson on May 28, 1833, Mr. Creed, one of the secretaries of the Railway, says — 'Nothing is said as to the appointment of engineer or solicitor, but I think *you* may be easy on that head. You have friends here and at Birmingham who appreciate your merits and services.' It was not, however, till just four months after the date of this letter that Robert Stephenson signed the contract that secured to him the post for which he had fought so zealously. In his note-book, under date September 20, 1833, is the following entry : — 'Signed contract with the London and Birmingham Railway directors, before Mr. Barker, at the Hummums, Covent Garden. Dined with Stanhope directors.'

On receiving the appointment of engineer-in-chief to the London and Birmingham Railway, Robert Stephenson broke up his modest establishment in Greenfield Place and took a comfortable house on Haverstock Hill, Hampstead Road, where he continued to reside for many years. From this time London became his home, and though he frequently visited Newcastle (the spirit of which enterprising and noble town had contributed greatly to form his character) and continued till his death to superintend the affairs of the engine manufactory, he never again had a home on the banks of the Tyne.

His residence in Newcastle had been broken by repeated periods of absence, during which he superintended works for his father, made trips to London and the continent, constructed the Warrington line, and the Leicester and Swannington Railway, surveyed the route for the London and Birmingham line, and directed the first operations at the Snibston colliery. These periods of absence reduce the time of his Newcastle life to a comparatively short term.

Still into that brief space much work and happiness had been compressed.

Numerous engagements left him little time for society. His domestic life, therefore, was strictly private, only three or four close friends being admitted to his house. One of those few intimate associates still lives to recall the happy evenings they occasionally spent in Greenfield Place, with music, talk, and cigars.

To these evening parties the pupils at the works were frequently invited. To limit the number of these pupils it was soon found necessary to raise the premium. Even at the increased rate there were found too many candidates for admission to 'the works;' and Robert Stephenson, whose sense of duty would not allow him to pocket a premium and give just nothing in return for it, resolutely declined to receive more than such a number of pupils as he conscientiously believed would profit by the opportunities offered them of acquiring information.

Inasmuch (Robert wrote from Dieppe, July 11, 1833, to his partner Mr. Richardson) as my own feelings are concerned, I should have no objections to receiving another apprentice into our establishment. The objections that exist are these. We have at present as many, indeed more, young men than we can sufficiently employ. If we increase the number (which we have very frequent opportunities of doing) we should only be doing the young men injustice, because they would not have proper and sufficient experience to learn the profession. They would be inadequately employed, and would consequently contract habits not calculated to advance them in after life. We are at present under an engagement to take a friend of Mr. Lock's (the Marquis of Stafford's agent), and when he comes our office will be really too full, even when I look forward to the London and Birmingham Railway going on. Taking young men, although it may be a profitable part of our business, is one that incurs great responsibility, which we feel is now as great as it ought to be. If these objections had not existed, it would have afforded both my father and self very great pleasure to take any young man introduced by you.

One of the pleasant features of Robert Stephenson's career was the strong personal attachment he formed for his pupils when they were young men of capacity and character. He never forgot or lost sight of them. A pupil of the 'right sort' was sure to win his approval and notice, and the pupil who had so earned his good opinion was sure to reap advantage from it. On the other hand, Robert Stephenson never considered himself either bound, or at liberty, to recommend for advancement an old apprentice, when he could not do so *honestly*. 'I can do nothing for you, unless you like to stop here as an ordinary workman,' he said to more than one pupil when his time was out: but then the young men to whom he so spoke merited no other treatment.

CHAPTER X.

CONSTRUCTION OF THE LONDON AND BIRMINGHAM RAILWAY.

(ÆTAT 29-34.)

Appointment as Engineer-in-Chief to the London and Birmingham Line — Contract Plans — Drawing-Office in the Cottage on the Edgeware Road, and subsequently at the Eyre Arms, St. John's Wood — Health and Habits of Life — Staff of Assistant and Sub-Assistant Engineers — The principal Contractors — Primrose Hill Tunnel — Blisworth Cutting — Wolverton Embankment and Viaduct — Kilsby Tunnel — Interview with Dr. Arnold at Rugby — Conduct and Character of Navvies — Anecdotes — Robert Stephenson proposes the Extension of the Line from Camden Town to Euston Square — Proposition first rejected and then adopted by Directors — Act of Parliament obtained for Extension of the Line — The Incline from Camden Town to Euston Square originally worked by Stationary Engines and Ropes — Lieut. Lecount's Comparison of Labour expended on the London and Birmingham Railway, and Labour expended on the Great Pyramid — Conduct of a certain Section of the Directors to Robert Stephenson — Opening of the Line — Dinner at Dee's Royal Hotel, Manchester — Robert Stephenson's Anger with a Director — Dinner and Testimonial given to Robert Stephenson at Dunchurch — Brunel uses Robert Stephenson's System of Drawing on the Great Western — Robert Stephenson's Appointment as Consulting Engineer.

THE labours of three surveys having been accomplished, the inordinate demands of landholders of every rank and condition having been satisfied, and a defeat as iniquitous on the part of the conquerors as any to be found in the chronicles of parliamentary warfare having been sustained, the London and Birmingham Railway Company had at length obtained their Bill. They had gained their

point on a new trial : but when Parliament reverses the
unjust decision of a preceding session, the injured party
has still to pay the costs of previous injustice.   The sum
of £72,869 recorded in the Company's books as paid for
obtaining their Act of Incorporation is an eloquent memorial of a conflict that stirred Westminster thirty years
since.

The Bill however was won, the Royal assent being
granted on May 6, 1833.   Mr. Isaac Solly, the first
chairman, was succeeded in 1834 by Mr. George Carr
Glyn, M.P., under whose able direction the line was
completed, and was brought to its present high state of
prosperity.   The appointment of an engineer was the
next affair for consideration.   Three years' indefatigable
attention to the interests of the Company gave Robert
Stephenson a claim upon their gratitude.   His display
of capacity during successive examinations before parliamentary committees had raised him high in the esteem
of his profession and the public.   A strong party, composed principally of his father's Liverpool antagonists,
spared no pains, however, to snatch from him the appointment of engineer-in-chief.   'He is a promising
young man, but still he is only a *young man*,' these
gentlemen repeated in every quarter, forgetting that public
railways were young things, and that the men best
qualified to construct the new roads were all young men—
the pupils of George Stephenson, who was himself still in
the middle period of life.

Fortunately Robert Stephenson's enemies were borne
down by more prudent and more honest directors; and
on September 7, 1833, the board resolved — 'That Mr.
Robert Stephenson be appointed engineer-in-chief for the

whole line at a salary of £1,500 per annum, and an addition of £200 per annum to cover all contingent expenses,
subject to the rules and regulations for the engineers'
department, as approved by the respective committees.' *
On Mr. Brunel's appointment as engineer to lay down
the Great Western Railway, with an annual stipend of
£2,000, Robert Stephenson's smaller salary was increased
to the same amount, the directors of the London and
Birmingham line rightly thinking that their character was
concerned in treating their engineer not less liberally than
Brunel was treated by a similar association.

In their next published report, dated September 19,
1833, the directors thus speak of their engineer's appointment —' The directors, considering it indispensable that,
in the execution of the works, one engineer should have
entire direction, and that his time and services should be
exclusively devoted to the Company, have under these conditions appointed Mr. Robert Stephenson engineer-in-chief
for the whole line ; and they are persuaded that to no one
could this charge be more safely or more properly confided.
He has received instructions to stake out the line without
delay, and the directors have reasons to expect that the
railway will be completed in about four years from the
commencement of the work.'

Having at length secured the post, Robert Stephenson
quitted Newcastle and came to the scene of his next five
years' labour. For a short time he resided in a furnished
cottage in St. John's Wood; but as soon as it was fitted

---

* The above resolution was, for
the purpose of this work, extracted
from the Minutes of the London
and Birmingham Railway, by the
late Admiral Moorsom, R.N., who
at the time of his lamented death
was chairman of the Company, to
which at its first outset he acted as
secretary.

up and ready for his reception, he moved into the house
on Haverstock Hill, which continued to be his home as
long as his wife lived.

He had undertaken a stupendous task. Up to that
time no railway of similar magnitude had been attempted.
The line from Liverpool to Manchester was by comparison
a trifling work. Its length was little more than a quarter
of the length of the new road, and its most important
works, including the Sankey viaduct (with nine arches
each of fifty feet span thrown over the Sankey valley, and
running seventy feet above the Sankey canal), its principal
tunnel, 2,250 yards long, and its firm highway over the
bogs of Parr Moss and Chat Moss, are in respect of magnitude not to be compared with the Kilsby tunnel, the
Blisworth cutting, and the Wolverton embankment and
viaduct.

A man of iron nerve would have experienced some uneasiness at the commencement of such an undertaking.
But Robert Stephenson, unlike his father, had throughout
life to contend with a distrust of himself, which was
partly due to innate modesty of disposition, and partly
attributable to a delicate nervous organisation. Though
the climate of South America had saved him from pulmonary consumption, he had by no means acquired the
soundness of constitution which young men ordinarily
enjoy. He was never a really strong man; and the
exertions of the four preceding years brought him to
London in 1833 in a very unsatisfactory condition of
health.

Had circumstances left him free to follow his own incli-
nations, Robert Stephenson, instead of taking a conspicuous
position in London society, would have passed his whole

life at Newcastle in comparative retirement. Naturally
no man was more averse to the turmoil of public life ;
no man more prized the tranquillity of home. He had
also become intensely fond of the mechanical part of his
profession. His labours in the Newcastle factory had been
attended with so much genuine pleasure, that he did
not without reluctance give them up for a more ambitious career; and in his later years he repeatedly declared to his intimate companions the regret he felt at not
having remained at Newcastle as a builder of locomotives,
though he had risen to be the most successful civil engineer of his time.

The engineer wished to ascertain with accuracy the
amount of the work before him. To effect this, before
cutting a turf, he went over every inch of ground, and
endeavoured to calculate the exact cost of every operation necessary for the accomplishment of his task. Hitherto, in laying down railways, engineers had been accustomed to do their work piecemeal, making a commencement, working up to difficulties, and then seeing how those
difficulties should be overcome. In laying down the Liverpool and Manchester Railway, George Stephenson had
at the outset of the undertaking only a general notion of
the labour before him. The details were not considered
till their consideration could no longer be deferred.
Robert Stephenson saw that this plan of leaving each day
to take care of its own evils was little calculated for
so vast an undertaking as the London and Birmingham
line. If the 112 miles of the proposed railway between
Camden Town and Birmingham were to be completed
within four or five years, the works must be advanced at
various points simultaneously, and the engineer-in-chief

must, at their commencement, have an accurate knowledge of their minutest details.

Robert Stephenson also resolved on making plans of
every part of the entire line, with unprecedented minute-
ness and completeness of detail.   He not only had a full
survey made, showing every natural feature of the route,
but prepared complete drawings for every work that
was to be executed, in all its details, accompanied with
full descriptions and specifications and accurate calculations of all the labour and material it would require.   As
each portion of the line was thus mapped out it was let
to a contractor, who engaged to complete the work for
a certain sum, and at the same time specified the exact
sum charged for each portion of the contract.   In those
days there were no gigantic contractors, a contract for
£100,000 being regarded as very large.   Men who in
he course of a few years made enormous fortunes were
then modest speculators, and had not sufficient funds in
hand to keep a regiment of 'navvies' at work for more
than a month.   The first contractors on the London and
Birmingham line were paid monthly, and in facilitating
these monthly payments the accuracy of the contract plans
was of the greatest service.   As the end of each month
came round, the assistant-engineer appointed over each
division of the line marked out the exact quantity of
work each contractor had accomplished, and for that
quantity payment was made.

It is difficult to give the reader any adequate idea of
the labour expended on these plans; for they had not
only to be made with the greatest attention to accuracy,
every separate calculation relating to them being three
or four times verified, but when they were made they had

to be multiplied.   The original contract drawings, signed
by the engineer-in-chief and the contractor, were preserved as documents of legal testimony; and of each of
them three copies were made — one for the use of the
committee, one for the engineer-in-chief, and one for the
assistant-engineer superintending the district in which the
work was situated. The entire line, as far as contracts were
concerned, was divided into thirty separate divisions, each
requiring distinct drawings, estimates, and specifications.
All these works, with two or three unimportant exceptions, were let to various contractors between May 1834
and October 1835.   From these data it may be seen
that the demands on Robert Stephenson's drawing establishment were very heavy.   It was calculated that, for
eighteen months, as many 'as thirty drawings per week,
each requiring two days' work from one pair of hands,
were turned out from the engineer-in-chief's office.'

Robert Stephenson was fortunate in having good subordinates.   Reserving a district, extending nine miles from
Maiden Lane, Camden Town, for his own especial supervision, he divided the remaining 103 miles into four
districts, each district having an assistant-engineer to
superintend it, and each assistant-engineer being supported
by a staff of three sub-assistants. For purposes of construction the line was thus apportioned—

District No. I.

This district, reserved for the engineer-in-chief's especial
personal supervision, extended from Camden Town for about
nine miles, and on its completion comprised the Camden Town
station, the Primrose Hill tunnel, the tunnel under Kensal
Green, and the bridge over the River Brent. The principal
engineer of this district, under Mr. Stephenson, was John

Birkinshaw, who was assisted by Mortimer Young, whose place was subsequently filled by Timothy Jenkins.

District No. II.

Assistant-engineer G. W. Buck; sub-assistant engineers, Mr., now Sir J. Charles Fox, F. Young, and Capt. Cleather, R.S.C. This district, extending from Harrow to Tring (23 miles) concluded with the Watford tunnel.

District No. III.

Assistant-engineer, John Crossley; sub-assistant engineers, S. S. Bennett, E. Jackson, J. Gandell, and M. Farrell. This district, extending from Tring to Wolverton (22 miles), included the Tring cutting and the Wolverton viaduct.

District No. IV.

Assistant-engineer, Frank Forster, who (on his succeeding to the post of assistant-engineer of District No. V.) was succeeded by G. H. Phipps; sub-assistant engineers, H. Lee, E. Dixon, C. Lean, and J. Brunton. This district, reaching over Wolverton and Kilsby (24 miles), included the Kilsby tunnel.

District No. V.

Assistant-engineer, Thomas Longridge Gooch, who (on his appointment to be the chief-engineer of the Manchester and Leeds Railway) was succeeded by Frank Forster; sub-assistant engineers, John Reid, B. L. Dickenson, M. Monteleagre, R. B. Dockray, and Lieut. P. Lecount, R.N. Extending from Kilsby to Birmingham : this district had for its principal works the Avon and Lawley Street viaducts.

The foregoing table assigns more than three sub-assistant engineers to the three last districts. There were, however, only three sub-assistants acting on any one district at the same time.

Robert Stephenson's first drawing office, whilst he was preparing the contract plans, was a small cottage standing on land which the Company purchased, near the point where the railroad passes under the Edgeware Road. This modest tenement was soon found to be too small for the

engineer's purpose. Luckily the Eyre Arms Hotel, St.
John's Wood, was just then vacant. The Company hired
it for their engineer's use, and 'the great room,' familiar
to many of the London public as a place of assemblage
for lectures, soirées, and political business, was speedily
furnished with drawing-tables and peopled with between
twenty and thirty draughtsmen. Amongst the gentlemen
employed at the Eyre Arms was Mr. G. P. Bidder, who
recently filled the office of President of the Institution
of Civil Engineers.

Eventually the line was let out in the manner indicated
by the following table:—

Name of Contract	Original Contractors	Date	Second Contractor
Euston Extension	W. and L. Cubitt	Dec. 1835	
Primrose Hill	Jackson and Sheddon	May 1834	The Company, Nov. 1834
Harrow	Nowell and Sons	May 1834	
Watford	Copeland and Harding	May 1834	
King's Langley	W. and L. Cubitt	Sept. 1835	
Berkhamstead	W. and L. Cubitt	Sept. 1835	
Aldbury	W. and L. Cubitt	Sept. 1835	
Tring	T. Townsend	Sept. 1834	The Company, Oct. 1837
Leighton Buzzard	James Nowell	Sept. 1835	
Stoke Hammond	E. W. Norris	Sept. 1835	
Bletchley	John Burge	Sept. 1835	
Wolverton	William Soars	Oct. 1834	The Company, June 1837
Wolverton Viaduct	James Nowell	Feb. 1835	
Castlethorpe	William Soars	Oct. 1834	Craven & Sons, July 1835
Blisworth	William Hughes	Feb. 1835	The Company, Dec. 1836
Bugbrook	John Chapman	Feb. 1835	
Stowe Hill	John Chapman	Feb. 1835	
Weedon	Edward Boddington	May 1835	W. and J. Simmons, May 1836
Brock Hall	J. and G. Thornton	May 1835	
Long Buckby	J. and G. Thornton	May 1835	
Kilsby	Jos. Nowell and Sons	May 1835	The Company, Feb. 1836
Rugby	Samuel Hemming	Nov. 1835	The Company, Oct. 1837
Long Lawford	W. and J. Simmons	Feb. 1835	
Brandon	Samuel Hemming	Feb. 1835	The Company, Jan. 1838
Avon Viaduct	Samuel Hemming	Nov. 1835	
Coventry	Greenshields and Cudd	Nov. 1834	The Company, May 1837
Berkswell	Daniel Pritchard	Nov. 1834	
Yardley	Joseph Thornton	Aug. 1834	
Saltley	James Diggle	Aug. 1834	
Rea Viaduct	James Nowell	Aug. 1834	

In this table may be seen the fortune attending the engagements of several contractors. The chief contracts —those, namely, for the tunnel at Primrose Hill, the Kilsby tunnel, and the Blisworth cutting—returned to the hands of the Company unfinished, and were perfected by the Company without the intervention of contractors : and in addition to these larger works, numerous smaller operations were beyond the powers of the commercial agents. It is not difficult to account for this collapse of contractors. Railway enterprise was still only in its infancy, and, though allowance had been made in estimates and contractors' agreements for a large rise in the price of labour, iron, and other materials, that allowance fell far short of the enormous and rapid advances made in the value of those commodities. Again, railway work was new, and the engineers were scarcely more prepared than the contractors for some of the difficulties with which they had to contend.

The Primrose Hill tunnel was one case of unexpected difficulty. The tunnel, passing under the high ridge between Hampstead and Primrose Hill, near Chalk Farm, is driven through a formation of blue clay, the extreme mobility of which, on exposure to moisture, offers peculiar difficulties to engineers. Years before the construction of the London and Birmingham line an attempt to drive a tunnel through this formation had terminated in failure, in consequence of the clay bearing down the brickwork. Warned by this case, Robert Stephenson proceeded at Primrose Hill with the greatest caution. As soon as a length of about nine feet of the excavation was finished, that portion of the tunnel was supported with strong timbers, and carefully lined with brickwork in mortar before any more earth was removed. Even this care, however, was insufficient. The pressure of the clay first forced out the mortar from the joints, and then crushed the bricks of the arch. To meet this difficulty, Robert Stephenson used only the hardest possible bricks, and laid them with Roman cement instead of mortar. This cement dries and becomes hard much sooner than mortar. The consequence of this change of material was the construction of a firm and durable lining of brickwork before the weight of the clay above was able to break in the walls of the passage. The experiment having proved successful, Robert Stephenson made himself doubly secure by making the brickwork much thicker than the estimates proposed it should be. In some portions of the Primrose Hill tunnel the thickness of the brickwork is only eighteen inches, but the larger portion is laid with a thickness of twenty-seven inches. And throughout the work costly Roman cement is used. No reader of these particulars will be surprised to learn that the difference between the estimated cost and the eventual cost of the tunnel was £160,000. The Primrose Hill contract was let for £120,000; it was not accomplished without an outlay of £280,000. No wonder, therefore, that the Company had to take back the work from the contractors unfinished.

Again, the operations of the Blisworth cutting exceeded the limits of the estimates so far that there was no prospect of their completion until the Company parted with their contractor. This excavation, which according to the estimate was to have contained 800,000 cubic yards, was not finished till nearly 1,000,000 cubic yards of earth and rock had been removed. At this

point of line 700 or 800 men, under the immediate command of the assistant-engineer, Mr. Phipps, were for
many months continually employed. For blasting the
limestone, there was for some time a weekly consumption
of 2,500 lbs. of gunpowder.

The Wolverton embankment, another of the contracts
which came back to the Company for completion, gave
the engineer much anxiety. In an embankment a mile
and a half long, exclusive of the Wolverton viaduct,
some difficulty was anticipated ; but human foresight
could not have provided for all the disasters attending
its construction. The embankment on the north side
of the viaduct gave comparatively little trouble. Composed of blue clay, lias, limestone, gravel, and sand,
it stood well, except at one place where it slipped, not
from its own weakness, but because the ground gave
way beneath its enormous weight. On the south side
of the viaduct, however, a grievous mishap occurred,
in the form of 'a slip,' that was not overcome for
months. No sooner was the way seen how to fill up
the slip, than Robert Stephenson was informed that the
troublesome embankment had caught fire. In its composition was a portion of alum shale, containing sulphuret
of iron. This material decomposing afforded a striking
instance of spontaneous combustion. Great was the consternation of the peasants at beholding a railway on fire.
Roguery was, they were convinced, at the bottom of the
catastrophe.

This same embankment was also the cause of difficulty
and litigation, which must be detailed at some length.

It would be a mistake to suppose that, with the passing of their Act, there was an end of vexatious opposition

to the London and Birmingham Railway Company. Beaten in Parliament, in a great measure through their bribery being exceeded by the bribery of their opponents, the persons interested in the Grand Junction Canal would not consent to relinquish the fight without another struggle.

The 85th, 86th, and 87th sections of the Act had reference to the rights and privileges of the Grand Junction Canal, over which the London and Birmingham Company proposed to carry their railroad in the parish of Wolverton. The 85th section provided that—

Nothing in the Act contained should diminish, alter, prejudice, affect, or take away any of the rights, privileges, powers, or authorities, vested in the Company of proprietors of the Grand Junction Canal, or authorise or empower the plaintiffs to alter the line or level of the canal or towing-path thereto, or any part thereof, or to obstruct the navigation of the canal or towing-path thereto, or any part thereof, or to obstruct the navigation of the canal, or any part thereof, or to divert the waters therein, or which supply the canal, or to injure any of the works thereof, and that it should not be lawful for the plaintiffs to make any deviation from the course or direction of the railway, as delineated in the maps or plans.*

With regard to the bridges which the Railway Company was empowered to make over the canal, the 86th section enacted that they should be —

Good and substantial bridges over the canal and the towing-path thereto, with proper approaches to each such bridge, and the soffit of each such bridge should be at least ten feet above the top-water level of the canal at the centre of the water-way, and no part of the arch over the towing-path should be less than eight feet above the top-water level of the canal, and each such bridge should be of such width and curve as should leave

---

* Railway and Canal Cases. Nicholl, Hare, and Carrow's Reports, vol. i. p. 224.

a clear, uniform, and uninterrupted opening of not less than
twenty-two feet for the water-way, and eight feet for the towing-
path under each bridge.

The Railway Company was also

Required during the progress of constructing each such
bridge over the canal, and of the necessary repairs or removal
thereof, from time to time, and at all times, to leave an open
and uninterrupted navigable water-way in the canal of not less
than sixteen feet in width, during the time of constructing and
putting in the foundation walls of the abutment of each of the
bridges, and of the new towing-path along the same, up to one
foot above the top-water level of the canal, and which time
should not exceed fifteen days; nor should less than twenty-two
feet for the water-way, and eight feet for the towing-path, be
left during the remainder of the period of constructing or repairing or removing each such bridge, and that the then present
towing-path should remain undisturbed until the new towing-
path wall should be erected, and the grounds made good and
properly gravelled and open for the free passage of horses
under each bridge.

The 87th section fixed certain penalties to be paid by
the Railway Company, and specified the manner in which
the Canal Company might recover such penalties, in case
any of the provisions of sections 85 and 86 should be
neglected.   Such were the precautions taken by the
Act to preserve uninjured the property of the Canal
Company.

The country at Wolverton, through which the London
and Birmingham line now runs, lies high upon the south
bank. Southward of the canal the railway passes through
extensive cuttings until it arrives within 150 yards of
the water.   At that point it enters upon an embankment which leads to the viaduct over the canal, and
extends 2,450 yards beyond it upon the northern side.
The entire embankment, comprising the small distance

on the south side and the large extent on the north,
contains 927,000 cubic yards of earth.   In order to construct the 2,450 yards of the northward embankment,
Robert Stephenson decided to convey 600,000 cubic
yards of earth across the canal from the many deep
cuttings in the southward country.   To convey this enormous quantity of earth across the water, it was necessary
to make a temporary passage of communication, the
construction of which involved the necessity of sinking
piles into the bed of the canal.   In the December of 1834
the embankment on the south bank had been carried
within twenty yards of the water, and it was time to
commence the embankment on the opposite side.
Robert, therefore, took his preliminary steps for constructing the temporary bridge.   At this juncture the Canal
Company intimated that the Act did not empower the
railway engineer to interfere with the water way.
Thinking the best way to avoid a dispute was by prompt
action, to change the discussion on what he *might do* into
a discussion on what he *had done*, Robert Stephenson
concentrated a strong body of engineers and navvies at
Wolverton, and without advertising his proceedings in the
papers or sending a notice of them to the office of the
Canal Company, proceeded to drive piles on the night of
December 23.   Relays of men carried on the work
without intermission by day-light and torch-light.   The
piles were driven into the bed of the river ; other piles or
supports were driven into the land on the north side, for
the purpose of sustaining the bridge ; beams were laid
from the piles in the water to those on the north shore ;
and by noon on December 25 (the toil having been
carried on through Christmas Eve into Christmas Day) the

temporary bridge was completed. The indignation of the
Canal Company at such a desecration of Christmas Day
may be imagined. Forthwith the directors of the powerful interest held counsel, and the result of their delibera-
tions was that on December 30 Mr. Lake, their engineer,
and a strong party of workmen, proceeded to the bridge
(which had been carried over the canal in little more
than a day and a half) and removed the piles which
supported it.

The next step was a petition on the part of the Railway Company to the Court of Chancery to restrain
the Canal Company from interfering with the operations
of the said Railway Company, and particularly from
'putting down, taking up, or destroying all or any or
either of the works to be made by the plaintiffs, their
servants or workmen, *for the purpose of making, constructing, or otherwise hindering or preventing or delaying* the
plaintiffs in making and constructing a passage of communication over and across the canal at Wolverton
aforesaid, in order to construct and complete the before-
mentioned embankment, and for transporting, by means
of such communication, the earth and materials whereof
the same embankment is to consist, over and across the
canal,' the plaintiffs of course undertaking to observe all
the stipulations, conditions, and provisions, of the 85th
and 86th sections of their Act, so as not to injure the
property of the Canal Company.

The case was argued, January 19, 1835, before the
Master of the Rolls, Sir C. C. Pepys, Mr. Pemberton and
Mr. Bacon being in support of the motion, and Sir C.
Wetherell and Mr. Turner appearing on the other side.

For the Canal Company it was not contended that the

piles and works of the temporary passage injured in any
way the bed of the canal, obstructed navigation, or im-
peded the tow-paths. The defendants only maintained
that the Act gave the Railway Company no right to make
such bridge, and therefore they would not let the foundations of such temporary bridge be put in the bed of their
water-passage. It was true the 8th section of the Act
authorised the Railway Company to ' make or construct,
upon, across, under, or over the railway or other works,
or any lands, streets, hills, valleys, roads, railroads, or
tram-roads, rivers, *canals*, brooks, streams, or other
waters, such inclined planes, tunnels, embankments, aque-
ducts, *bridges*, roads, ways, passages, conduits, drains,
piers, arches, cuttings, and fences,' as they should think
proper for the purpose of carrying out their undertaking.
But it was maintained that the 85th and 86th sections
restricted the privileges granted by the 8th clause.

Of course the counsel in support of the prayer contended that, whereas the 8th clause authorised the plaintiffs
to construct any temporary bridge necessary for making
their line, the 85th and 86th clauses referred only to permanent and not temporary bridges, and therefore could
in no way be construed as qualifying the prior permission.
Much to the delight of Robert Stephenson, who sate in
court throughout the hearing of the cause, the Master of
the Rolls in a lucid and admirable judgement granted the
injunction.

But the most obstinate and costly of all the contests
involved in carrying out the works came off at the Kilsby
Tunnel, about six miles from the Rugby station. Robert
Stephenson's original plan was to lead his road from Birmingham to London by way of Northampton, but the inha-

bitants of Northampton raised so effectual an opposition to
the scheme, that the engineer was necessitated to choose a
route along which adverse influence was less powerful.

The consequence of the opposition was hurtful alike to
the town and the Company. The inhabitants of the town,
after repenting their folly, had to petition humbly for a
branch line, and the Company were driven to bore a way
for their rails through the Kilsby ridge at the stupendous
outlay of more than £320,000. The length of this costly
passage, situated about six miles on the London side of
the Rugby station, is just 2,400 yards. A few facts,
briefly stated, will enable the reader to form some conception of the labour expended upon it. Thirty-six
millions of bricks were used in its construction. The
two shafts by which it is ventilated and supplied with
light are sixty feet in diameter, and the deeper of
them contains above a million of bricks. These two
enormous shafts the walls of which are perpendicular,
were built from the top downwards, small portions of the
wall (from six to twelve feet long and ten feet deep) being
excavated at a time, and then bricked up with three feet
depth of bricks, laid with Roman cement. At one time
1,250 labourers were employed in building the tunnel.
To lodge and cater for this army of navvies, a town
of petty dealers soon sprung up ; sheds of rude and
unstable construction rose on the hill above the tunnel,
and in them a navvy could obtain at a high rent
the sixteenth part of a bed-room. Frequently one room
containing four beds was occupied by eight day and eight
night workmen, who slept two in a bed, and shifted
their tenancies like the heroes of a well-known farce.

The disasters of the Kilsby excavation were dimly

foreseen and predicted by Dr. Arnold.   On his first visit
to Rugby after the Bill for the London and Birmingham
Railway had received the Royal assent, Robert Stephenson
called on the great schoolmaster to offer him his respects.
The young man brought no letter of introduction, and
either was, or imagined himself to be, received with cold-
ness and hauteur.   Dr. Arnold was certainly polite, but
perhaps formal, his manners being of a school with
which, at that period of his life, Robert Stephenson was
not familiar.   Anyhow the interview left on the mind
of the engineer an unpleasant impression, which was
doubtless in some part due to Arnold's last words : ' Well,
sir,' he said, pointing in the direction of the Kilsby ridge,
' I understand you carry your line through those hills.
I confess I shall be much surprised if they do not give
you some trouble.'

In due course the trouble came.   Trial shafts sunk at
various points ascertained that the line of the proposed
tunnel ran for the most part through lias, shale, and beds
of rock with sand.   They proved also that in places there
would be a considerable quantity of water.   The difficulties
apprehended were not trivial; but Messrs. J. Nowell and
Sons felt that they could cope with them at an outlay
short of £99,000, and for that sum they undertook the
work.   It was not long before they had reason to repent
the bargain.   To afford exit for the soil removed, Robert
Stephenson ordered the sinking of eighteen working
shafts.   The second of these shafts came upon a bed of
gravel and sand, containing a great deal of water,
overlaid by forty feet of clay.   Repeated borings discovered this quicksand to be a basin, lying along one side
of the hill, and extending 400 yards over the line of

the tunnel.   As the evil fortune of Messrs. Nowell and Sons and their employers would have it, this treacherous basin had been missed by the trial shafts only by a few feet.   Ruin stared the contractors in the face, and Mr. Nowell, whose health had for some time been declining, died shortly after the discovery of the quicksand, his death being doubtless accelerated by the fulfillment of Dr. Arnold's prediction.

The calamity which had prostrated, if not killed, their principal contractor was not without its influence on the directors.   Amongst them were those who seized it as an occasion for insinuating that their 'young engineer' was at fault, and that, had he had more experience, the trial shafts would have discovered the dangerous spot.   The consternation of both committees (the London committee and that which sate at Birmingham) was at its height when Captain Moorsom, in his official capacity of secretary and business adviser, was deputed to visit Kilsby, hold an interview with Robert Stephenson, and urge upon him the propriety of calling in further engineering advice.   Without delay Captain Moorsom acted upon his instructions, and arriving at Kilsby, hastened to the office, where he found Robert Stephenson holding a consultation with his assistant and sub-assistant engineers.

When Captain Moorsom made his presence known, and stated with delicacy the anxiety of the directors, and the satisfaction they would feel in calling in other engineering advice, Robert Stephenson answered cordially and without irritation, 'No ; the time has not come for that yet.   I have decided what to do.   I mean to pump the water all out, and then drive the tunnel under the dry sand.   Tell the directors not to be frightened, and say that all I ask is time and fair play. If I can't get rid of the water, I'll then think about going to other engineers for help.'

Captain Moorsom then knew but little of Robert Stephenson. He had seen him occasionally in parliamentary committee-rooms, and had heard him spoken of by friends as a young man fortunate in the possession of extraordinary intellects — spoken of by enemies as a young man fortunate in the possession of an extraordinary father. From that time, however, Captain Moorsom became Robert Stephenson's enthusiastic supporter; and, returning to the directors, he told them to rest assured that their engineer deserved their entire confidence.

With the aid of 13 steam-engines, 200 horses, and 1,250 'navvies,' the engineer again set to work. A short distance from the line of the tunnel, shafts, cased with wooden tubbing, were forced through the sand, and from them headings were driven into the sand, through which the water flowed freely to the pumps. For nine months was the pumping continued, and for the principal part of that time each minute saw 1,800 gallons of water sucked from the basin. At length the difficulty was overcome. The tunnel was then shot under the sand, and the gentlemen who had anticipated the failure of their 'young engineer,' and who during the protracted trial had never ceased to worry him with impertinent criticisms, received a welcome and salutary lesson.

In November 1836, another trouble occurred in the irruption of an enormous body of water into a part of the tunnel where there were no pumps. The water rose rapidly, and (to save a portion of the tunnel) it was necessary forthwith to complete the lining of brickwork.

To effect this workmen were floated up the tunnel on a large raft; and, as fast as hands could move trowels and adjust bricks, the task was accomplished. Before it was completed, however, the water rose so high and with such increased rapidity, that the men on the raft were in danger of being jammed up against the roof of the tunnel. To save the party, Mr. Charles Lean, sub-assistant engineer, jumped into the water, and, swimming with a tow-line between his teeth, tugged his men to the foot of the nearest working shaft, through which they were drawn from their perilous position ' to bank.'

When the reader bears in mind that the last few pages relate only to three or four out of thirty or forty contracts, and also remembers that great exertions were made to carry out all the contracts simultaneously, he will not be surprised at learning that ' the navvies ' in Robert Stephenson's army were numbered by thousands.

The original Act for the London and Birmingham Railway empowered the Company to make a line ' commencing on the west side of the high road leading from London to Hampstead, at or near to the first bridge westward of the lock on the Regent's Canal at Camden Town, in the parish of St. Pancras, in the county of Middlesex, and terminating at or near to certain gardens, called Nova Scotia Gardens, in the parish of Aston juxta Birmingham and Saint Martin Birmingham, in the county of Warwick.' At the time of the parliamentary contests the projectors thought it would be more prudent not to alarm the public mind with a proposal to carry their road nearer London. As it was, the timid were predicting all sorts of evil consequences from an iron-way, by which all the evil-doers of London could in a moment fly beyond the police. A consideration, however, that had yet more weight with the Company was Lord Southampton's opposition to their undertaking.

When their petition was rejected by the peers, Lord Southampton had been a principal cause of their defeat. His lordship owned much of the land between Camden Town and the streets of the capital, and it was under a strong conviction that his property would be prejudiced by the railway that he opposed the project. To conciliate this powerful enemy, the projectors determined to interfere as little as possible with his estate. Scarcely, however, had the line been begun, when Lord Southampton began to entertain different views with regard to railways. The success of George Stephenson's lines, the Stockton and Darlington and the Liverpool and Manchester, was admitted to be beyond a doubt. The value of land adjacent to them had everywhere increased, in some places had increased enormously. London residents began to see that it would be to their interest to get the London and Birmingham terminus as near them as possible; and Lord Southampton perceived that the extension of the line through his estate would greatly increase its value.

Robert Stephenson was the first to detect the change in public feeling, and to suggest to the directors the advisability of getting another Act of Parliament, empowering them to carry their line to Lancaster Place, Strand, abutting on the Thames. Nervous and retiring, he could not get up courage to proffer this advice until he had talked the matter over many times with Mr. Charles Parker, the solicitor of the Company, and his own intimate and valued friend. Mr. Parker rallied him for being 'afraid of the board,' and urged that it was his duty to tell the Company what he honestly believed would promote their interests. In consequence of Mr. Parker's repeated exhortations Robert Stephenson laid his views before the directory, and for so doing was rewarded with an emphatic and almost unanimous snubbing by the gentlemen assembled, who feared to take so bold a step. He was told that he was an engineer; and it would be more becoming in him, as an engineer, to confine his attention to the matters of his profession, and not to concern himself with the affairs of others. Indignant, and for the moment humiliated, Robert Stephenson hastened to Mr. Parker, and communicated the result of his Quixotic attempt to benefit the Company. Again his friend rallied him, and, laughing at his mortification, told him that before the next meeting of the committee his suggestions would have favour with those same directors who had displayed such want of courtesy. The solicitor was no bad judge of the question and the men. Before many weeks had passed Robert Stephenson's scheme was supported both by the London and the Birmingham committee, and more especially urged forward by Mr. Wilson, the agent for Lord Southampton. In due course a new Act empowered the Company to extend their line, 'commencing in a field on the west side of the high-road leading from London to Hampstead, being the site of the depôt or station intended to be made for the use of the said railway, in the parish of St. Pancras, in the county of Middlesex, and thence passing across the Regent's Canal, between the first and second bridge westward of the lock at Camden Town, into and through the said parish of St. Pancras, and terminating in a vacant piece of ground in a place called

Euston Square, on the north side of Drummond Street, near Euston Square, in the same parish.' Thus part of the engineer's scheme was adopted. If the whole design had been approved, Robert Stephenson would have had the further credit of originating the system which has extended the lines across and through the metropolis.

Euston Square lies much lower than Camden Town; and the portion of the railway that lies between those points was worked for some years by ropes and stationary engines, on account of the steepness of the incline, and for no other reason. The trains from Euston Square were drawn up the incline at the rate of twenty miles an hour by an apparatus consisting of 10,000 feet of rope (six inches in circumference) and two stationary engines. These engines and their ropes cost £25,000. The up-trains were disjoined from the locomotives at Camden Town, and were carried down the inclination by gravity alone into the Euston station, and were prevented from attaining too great speed by the use of powerful brakes. The line between Euston Square and Camden Town was thus worked till the July of 1844, in which month loco-motives were employed to draw the laden carriages up the incline.* It may interest some readers to know that

---

* The late Admiral Moorsom, R.N., amongst other papers supplied by him for the biography of his friend, furnished the following extract from the Minutes of the London and Birmingham Line: —

'Friday: July 12, 1844.

'On and after Monday next the use of the rope will be wholly dis-continued, and all the trains taken from Euston by the locomotive en-gines.

'It will be necessary to notify to the locomotive department at Cam-den the weight of the engines, thus—

'When likely to be 16 carriages, one signal about 8 minutes before the time of departure.

'If likely to be 21, *one* signal 8 or 10 minutes, and a second 4 or 5 minutes before the time.

'(Signed)   H. P. Bruyeres.'

the stationary engines, discarded from Camden Town, are at the present time doing duty in a silver mine in Russia.

Thus Robert Stephenson and the army under his command began and completed in less than four years and three months the first metropolitan railway that was worked by locomotives. The first sod was cut at Chalk Farm on June 1, 1834, and the line was opened on September 15, 1838. On an average 12,000 men were throughout that space of time employed upon the works, i.e. rather more than 107 men to each mile. Estimating the labour expended upon the vast operations of these 12,000 men, Lieut. Lecount, R.N., one of the assistant engineers of the line, says —

The great Pyramid of Egypt, that stupendous monument which seems likely to exist to the end of all time, will afford a comparison. After making the necessary allowances for the foundations, galleries, &c., and reducing the whole to one uniform denomination, it will be found that the labour expended on the great pyramid was equivalent to lifting fifteen thousand seven hundred and sixty-three million cubic feet of stone one foot high. This labour was performed, according to Diodorus Siculus, by three hundred thousand, and according to Herodotus, by one hundred thousand, men, and it required for its execution twenty years. If we reduce in the same manner the labour expended in constructing the London and Birmingham Railway to one common denomination, the result is twenty-five thousand million cubic feet of material (reduced to the same weight as that used in constructing the pyramid) lifted one foot high, or nine thousand two hundred and sixty-seven million cubic feet more than were lifted one foot high in the construction of the pyramid. Yet this immense undertaking has been performed by about twenty thousand men in less than five years.

The reader will observe that Lieut. Lecount in making his calculation takes, not the average number of workmen employed on the line, but the highest number acting together at a time of special exertion.

It should be borne in mind that throughout this period, although the majority of the Directors did him full justice for integrity and talent, yet Robert Stephenson was harassed with the vexatious opposition of a section of those directors whom he was so zealously serving. It would do no good at this date to rake up the animosities of a generation fast disappearing from the world; but it is right, for the consolation and encouragement of honest men suffering under similar persecution, to publish the fact that, in addition to the anxiety and toil imposed upon him by his responsible position, he had to endure ungenerous treatment from his employers.

At length, after innumerable delays and an enormous excess of expenditure beyond the estimates, the line was opened with suitable, but modest, ceremony. The Committee of London Directors, accompanied by the principal officers of the line and a few friends, made a trip in one train to Birmingham and dined with the Birmingham committee at Dee's Royal Hotel, Robert Stephenson taking charge of the engine during the excursion.

To him the day was far from being a day of pleasure. In bidding adieu to a work magnificently completed, which had taken up several of the best years of his life, he felt that sadness which Gibbon experienced whilst penning the last lines of his history. To this depression was added the irritation of an insult offered to his father by one of his own principal enemies. That very morning, before mounting the engine to drive to Birmingham, Robert Stephenson had read in a newspaper an article full of base insinuations against, and reflections upon, his father.

In the evening a party of about one hundred people
assembled at Dee's Royal Hotel.   The banquet passed off
heavily, and on the following morning Robert Stephenson
met, after breakfast, the person who was supposed to be
the author of the article which had caused him so much
pain, and immediately asked him whether he had written
it.   The charge was admitted; and Robert Stephenson,
having expressed in the strongest terms his opinion on the
subject, left the room.   The writer of the article, who was
also a director of the Company, appealed for protection to
Mr. Glynn, the chairman, who was not present at the scene.
The latter replied briefly that if directors chose to attack
the engineer of the Company or his father in the public
journals, they must do so in their private capacity and at
their own risk.   Some years afterwards the director met
Robert Stephenson on the station platform at Rugby, and,
expressing his regret for the old quarrel, extended his
hand to the engineer, who instantly accepted it, and the
feud was forgotten.

A more agreeable celebration of the successful conclusion of the London and Birmingham line was a dinner
given to Robert Stephenson towards the close of the previous year (December 23, 1837), at Dunchurch in War-
wickshire, when the acting and assistant engineers presented the engineer-in-chief with a silver soup-tureen and
stand, worth 130 guineas, as an expression of their affectionate admiration.   Mr. Frank Forster was in the chair,
and Lieut. Lecount, R.N., the historian of 'the works,' in
the vice-chair.   George Stephenson was present as a
guest.   The host of the 'Dun Cow,' Dunchurch, had
never before entertained so distinguished a party.

An anecdote connected with the 'Dun Cow' dinner

must not be omitted. The subscription for the soup-tureen and stand was confined to the engineering officers of the Company — a restriction which excluded several persons who were anxious to subscribe. Mr. Charles Capper, who, having merely supplied a quantity of machinery to the line, could only be regarded as a sub-contractor, in vain endeavoured to force his contribution on the committee, who declined to accept it because, if they set aside 'the line' agreed upon, they should not know where to draw another. At the dinner, however, the enthusiastic sub-contractor was present in all his glory and admiration for Robert Stephenson. 'Anyhow,' he exclaimed to some of the committee, as he entered the room, 'you will allow me to dine with Mr. Stephenson.' As the dinner was public, there was of course no opposition. In the dining-room the testimonial was placed on a buffet for inspection; and as the guests assembled, they surrounded the soup-tureen and criticised it. At length the sub-contractor, with a glow of triumph in his face, exclaimed, 'It is a handsome tureen, but it wants a ladle.' And as the critic spoke, he supplied the deficiency by taking from his pocket a large and very handsome ladle, and putting it into the silver vessel.* The ladle formed part of the testimonial, and Robert Stephenson in after life was very proud to tell his friends how he became possessed of his large soup-ladle.

Thus was completed the construction of the London

---

* The following inscription was put on the tureen : —
'To Robert Stephenson, Esquire, Engineer-in-Chief of the London and Birmingham Railway, a tribute of respect and esteem from the members of the Engineering Department who were employed under him in the execution of that great work. Presented on the eve of their gradual separation.'

and Birmingham Railway, with which line Robert
Stephenson maintained his connection up to the time of
his death, acting as its consulting engineer with a salary
of £100 per annum, and his expenses when called to
attend on the line.   It was the first of our great metropolitan railroads, and its works are memorable examples
of engineering capacity.   They became a guide to succeeding engineers ; as also did the plans and drawings
with which the details of the undertaking were ' plotted '
in the Eyre Arms Hotel.   When Brunel entered upon
the construction of the Great Western line he borrowed
Robert Stephenson's plans, and used them as the best
possible system of draughting.   From that time they
became recognised models for railway practice.   To have
originated such plans and forms, thereby settling an
important division of engineering literature, would have
made a position for an ordinary man.   In the list of
Robert Stephenson's achievements such a service appears
so insignificant as scarcely to be worthy of note.

CHAPTER XI.

AFFAIRS, PUBLIC AND PRIVATE, DURING THE CONSTRUCTION OF THE LONDON AND BIRMINGHAM RAILWAY.

(ÆTAT. 29–35.)

Stanhope and Tyne Railway Company — Robert Stephenson appointed their Engineer — Opening of the Line and its rapidly increasing Embarrassments — Robert Stephenson visits Belgium with his Father —Offices in Duke Street, and George Street, Westminster —The Session of 1836 — Various proposed Lines between London and Brighton: Sir John Rennie's, Robert Stephenson's, Gibbs's, Cundy's—London and Blackwall Railway, and the Commercial Road Railway—Robert Stephenson strongly opposes the Use of Locomotives in Towns—Life at Haverstock Hill—Reading, Friends, Horses, Sunday Dinners — Newcastle Correspondence — Mrs. Stephenson's Accident to Knee-Cap—Professor Wheatstone's and Robert Stephenson's Adoption of the Electric Telegraph — Robert Stephenson assumes Arms — That ' Silly Picture.'

ALTHOUGH the terms of Robert Stephenson's agreement with the directors of the London and Birmingham Railway Company precluded him from undertaking the personal superintendence of any other engineering work during the construction of that line, he was at liberty to act as a consulting engineer in the civil department of his profession, to advise on questions of parliamentary tactics, to appear as a professional witness before committees, and to visit any part of the kingdom or continent, for brief periods — either to superintend the interests of his private undertakings, or inspect the scene of new public works. Haverstock Hill was his home ; and the course of the London and Birmingham line was the route on some part of which he might, on five days out of six, have been seen getting over rough ground on horseback—or walking from point to point, at such a pace that his companions, puffing at his heels, were frequently compelled to cry out for breathing time. But by careful distribution of his time he made leisure for many matters distinct from the first Metropolitan Railway.

His connection with the Stanhope and Tyne Railway had already become to him a source of serious uneasiness. As it for years caused him grave anxiety, and at one time threatened to plunge him in pecuniary embarrassment, it is fit here to speak at some length of that signal instance of rash speculation and grave mismanagement of amateur directors.

As early as 1831, a scheme was concocted by certain speculators to work some lime-quarries near the town of Stanhope, in the county of Durham, and certain portions of the extensive coal field at Medomsley, in the same county, and to connect the two works by a railway. The chance that such a line would answer was very slight; for the fifteen miles of rugged country through which it ran by a succession of unusually steep inclines was sparsely populated, and (for Durham) poor in minerals. A company was nevertheless formed, and the iron road was laid down. A few months' trial was sufficient to prove what ought to have been foreseen, that such a line could never pay. Two of the original projectors slipped out of the affair on profitable terms, leaving their companions to adopt a bold, and by no means unwise, suggestion, for making good their loss. The line from Stanhope to

Medomsley was a failure for obvious reasons; but it was argued that if the way were carried on twenty-four miles further, to South Shields at the mouth of the Tyne, it would pass through the heart of an extensive and productive coal field, and find abundance of business. This second scheme was just as sound as the original undertaking was bad; and had it only been carried out with prudence, it would have been eminently successful.

The new scheme immediately took, and the shares were subscribed for by people of credit, and in some cases of wealth, for the most part residing in London. The capital of the new Company was stated to be £150,000, consisting of 1,500 shares of £100 each. Of these shares, however, only 1,000 were ever paid upon, the remaining 500 being gratuitously allotted to the two projectors of the undertaking, who, in addition to this remuneration for their services, secured for themselves one half the profits of the line, after the proprietors had received 5 per cent. on their shares. Power was given to the directors to raise £50,000 more capital by the creation of new shares, and £150,000 on loan.

In the North of England it has been an ancient custom for speculators to lay down colliery tramways, without going through the tedious and costly process of parliamentary incorporation. Running from coal fields to neighbouring ports, these lines are never very long. As a general rule they run through the lands of but few owners, the value of whose property they enhance. It is therefore usual for projectors of such tramways to make their own agreements with landowners, paying a certain annual rent for right of way, or way-leave as it is called, and taking such way-leave for ninety-nine years, with

a reserved power to abandon on giving twelve months'
notice. The Stanhope and Tyne line was made on this
plan; but so badly were the negotiations with landowners managed, that when the line (in all thirty-five miles
long) was completed it was burdened with a way-leave
rental of more than £300 per mile. This was bad. But
a far worse consequence of the arrangement was one
inseparable from the system above described. Having no
act of incorporation, an ordinary way-leave railway is a
simple partnership affair, in which every shareholder is a
partner. And that meant, in times prior to the Limited
Liability Act, that every shareholder in an ordinary way-
leave tram company was personally responsible for all the
liabilities of the company.

From first to last, method and business exactness were
neglected in the affairs of the Stanhope and Tyne line.
The new Company's deed of settlement was not executed
till February 1834, but the first way-leave agreements
were entered into with landowners in April 1832 ; and as
far as confused accounts can be trusted, it would appear
that nearly the whole of the capital was paid up and
expended, and heavy debts were incurred before the
execution of the deed. One of the first acts of the Company was to draw a bill of exchange ; and when the proprietors at length decided to dissolve the association, the
bills in circulation for which the Company were responsible amounted to £176,000.

In an evil hour for Robert Stephenson the directors of
the Stanhope and Tyne line agreed to consult him as an
engineer. At first he was well pleased with the summons.
The remuneration for the services required of him was to
be £1,000 ; but he was persuaded to accept in payment

of that sum ten shares in the Company's stock. At first, Robert Stephenson liked his £1,000 all the better for being in that form, since his own judgement, as well as the observations of bystanders, assured him that the new railway must eventually answer. He did not calculate with a foreknowledge that the undertaking would be mismanaged. And he was at the time ignorant of the difference between the legal positions of a shareholder in an incorporated railway, and of a shareholder in a line without an act of incorporation.

The ultimate fate of this ill-starred Company will not at present be set forth. It is, however, best to notice, at this point, the course of its affairs during the construction of the London and Birmingham Railway. At great outlay the directors built staiths, and purchased freehold and leasehold houses, buildings, wharves, and quays at South Shields ; and in the March of 1835, on the projection of the Durham Junction Railway, in which the proprietors of the Stanhope and Tyne deemed themselves deeply interested, the directors of the latter Company subscribed £40,000 out of £80,000 to be raised for the new line. For the most part these purchases and new engagements were based on good considerations, and were such, that if the pecuniary obligations consequent upon them had been originally made on a proper scale, and had then been met in a proper way, no objection could have been preferred against them. But not content with buying at exorbitant prices, the new Company started with the ruinous system of borrowing on bills, instead of raising from amongst themselves, or by the creation of new shares, the sums necessary for liquidating debts. The fact was, the directory lay in the hands of persons whose circumstances precluded any other system of raising money. From first to last an important department of the business of the directors was to raise money on accommodation bills on terms averaging 11 per cent. per annum. In June 1834, following the February in which their deed of settlement was executed, the directors obtained on mortgage £60,000 from the Alliance Assurance Company. The railway and collieries commenced working in September 1834, and by the end of the year the entire expenditure of the Company amounted to £226,485 17s. 0d.; of which amount £100,000 had been received from payments on the 1,000 paid-up shares, £60,000 had come from the Alliance Assurance mortgage, and the remaining £66,485 17s. 0d. had been raised on bills. Twenty-four thousand pounds were soon afterwards raised by debentures. Thus affairs began, and thus they went on. Loan was raised on loan, bill accepted after bill. Every month affairs looked worse ; so that in 1838, when the London and Birmingham line was opened, instead of finding himself the owner of £1,000 in a railway the shares of which were at a premium, Robert Stephenson found himself with ten shares in an affair that was throughout the money-market a byword for failure— shares which he would gladly have been able to throw into the sea, since they rendered him personally liable for an enormous sum of money. Thus was Robert Stephenson paid for engineering services. He had done good work, and as a reward for the service he found insolvency staring him in the face. It was a salutary lesson to him. Ever afterwards he resolutely refused to take the shares of any company in payment for work done. He took, indeed, thirty shares in the London and Birmingham previous to the Act being obtained; when the directors, finding great difficulty in getting the proportion of subscriptions required by standing orders, called on all their principal officers to put down their names for shares. But that was a different affair ; and moreover, he had not then the experience of the four succeeding years to guide him.   To speculation of all sorts he had a dislike amounting to repugnance.   His investments gave modest dividends, but they were safe.   He believed in the maxim that a high rate of interest is only another name for bad security.   This distaste for pecuniary risk was seen in little things as well as great — in his amusements as well as his commercial arrangements.   He liked horse-races, and during the last years of his life always endeavoured to be at Epsom and at Ascot, but his most intimate friends never knew him to bet a shilling on any horse. In the same way, he enjoyed a rubber, but he never played for high stakes.

In the May of 1835, Robert Stephenson accompanied his father to Brussels; the elder and the younger engineer having been summoned by King Leopold to advise as to the construction of a complete system of railways for his kingdom.   On that occasion, when the father obtained the decoration of the Order of Leopold, the son was also admitted to familiar intercourse with the King.   Two years afterwards, on the public opening of the railway between Brussels and Ghent, when George Stephenson was received by the Belgians with an enthusiasm of admiration, Robert Stephenson renewed his acquaintance with a country which enjoys distinction amongst continental nations for an early and cordial adoption of railroad locomotion, and was again hospitably entertained by the sovereign, who in 1841 conferred on him, as he had six years before conferred on George Stephenson, the decoration of the Order of Leopold.

On his return from Belgium, Robert Stephenson found himself overwhelmed with work. The scant leisure left him by the London and Birmingham line was more than fully occupied with examining projects for new lines which sprung up in every direction, and concerning which his advice was sought alike by engineers and by the public. For two years he managed to attend to this extra-official business at his office at Camden Town, on the London and Birmingham works. In 1836, however, finding he could no longer, either with comfort to himself or with the approval of the Company, receive his daily levy of projectors and engineers at Camden Town, he took an office in Duke Street, Westminster. In the following year this office was relinquished for one in Great George Street, Westminster, with which street the Stephensons and their profession are intimately associated in the public mind. In that street Robert Stephenson, with the principal members of his staff, had offices up to the time of his death. On the doors of 24 Great George Street country sight-seers still read the name of the great constructor of railways and builder of bridges ; and in the adjoining mansion is established the Institution of Civil Engineers.

The years 1836 and 1837 were remarkable for railway enterprise. In the thick of the parliamentary fight Robert Stephenson appeared as professional witness, and more especially as the projector and engineer of a line between London and Brighton, which unfortunately miscarried, but was not shelved until it had engrossed a large amount of attention and discussion. As early as

1833, and indeed before that year, his attention was called to the subject of railway communication between the metropolis and the most fashionable watering-place of the country. In 1834 and in 1835, he was again consulted as to the lines projected between those points, and finding none of the proposed routes such as he could in all respects recommend he sketched out a line of his own. The consequence was that the session of 1836 saw four distinct applications to Parliament for different lines between London and the Sussex cliffs. The rival projects were Sir John Rennie's, or the direct line; Mr. Robert Stephenson's line; Mr. Joseph Gibbs's line; and Mr. Cundy's line. Here was a noble fight for Westminster, spoil for lawyers, agents, surveyors, and witnesses. Each of these proposed lines availed itself of a terminus already constructed, Stephenson's line taking the terminus of the London and Southampton Railway at Nine Elms, a little above Vauxhall, with a depôt on the banks of the Thames, and branching from the line at Wimbledon Common, five miles and a half from the terminus; and the direct line and Gibbs's both adopting the Greenwich Railway Terminus at London Bridge, and availing themselves of the railway, already sanctioned and under course of construction, as far as Croydon.

Mr. Cundy's projected line was a freak of daring such as can only be found in times of unusual excitement. In these comparatively sober days it can scarcely be believed that just five and twenty years ago a company should have contemplated the construction of an important line of railway, and, with full attention to all the costly forms of law, have applied to parliament for leave to construct their line, without having made any survey (in the engineering sense of the word) of the country through which the line was to pass, — having in fact trusted to the Ordnance map for knowledge of the natural features and levels of the ground. Mr. Walker, in his report on the project, drily remarked, 'The line of country, with the levels, Mr. Cundy shows would be very desirable and easy, *if they could be found*; but I have not succeeded in doing so, my levels being considerably different from Mr. Cundy's.' To well-informed engineers the bare mention of ' Cundy's line ' was a signal for hearty laughter. As soon as Robert Stephenson put forth his plan, Cundy fired up with virtuous indignation, and was shocked at the immorality of the engineer of the London and Birmingham Railway, who had pirated* his plans. No better illustration of the comic side of railway gambling a quarter of a century since could be made than a drama reproducing all the circumstances connected with ' Cundy's line.'

---

* 'If the Committee will permit me, I will just say one word with respect to Mr. Cundy's line, which had escaped me. Now Mr. Cundy's accusation against Mr. Stephenson is, that he has pirated his line. If he has, I do not know that it matters much, except with regard to Mr. Stephenson's character. If he had been a Red Rover amongst engineers, and had taken from another the fruit of his skill and labour, I do not think the Committee would attach much importance to it, as against the promoters of this Bill. But it is impossible that it can have been so ; for when Mr. Cundy comes to put in the Ordnance map, with his section of Mr. Stephenson's line marked upon it, it is found that he actually lays down this line a quarter of a mile distant from that on which he himself relies. Mr. Cundy has also made some very extravagant use of his engineering powers, *for he has made the River Mole ten feet higher than the stream which runs into it*; and he has told you a 74-gun ship may float upon the flood of that river, that you may go down there some Saturday night, that you may sleep there on Sunday night, and on the Monday morning you may see Mr. Stephenson, and his railroad, and his embankment gone, washed away by the flood.' — *Stephenson's London and Brighton Railway. Speech of the Hon. J. C. Talbot, on summing up the Engineering Evidence given in support of Stephenson's line before the Committee of the House of Commons.*

The contest, however, between Stephenson's line and
Rennie's and Gibbs's lines was honest, and involved many
delicate points of consideration. It was, however,
proved by the personal examination of the Committee,
that the alleged survey for the latter of these lines was
not much less incorrect than that of Cundy's line. Unable
to superintend the details of the survey himself, Robert
Stephenson committed them to the care of Mr. Bidder,
who, not without an appearance of justice, was regarded
by the public as being really and truly the engineer of
the line. As its name implies, the chief object of Sir John
Rennie's direct line was the shortest possible passage between the two termini, the natural obstacles of the country being boldly met, instead of being adroitly avoided.
Robert Stephenson and Mr. Bidder, on the contrary, aimed
at a line which, without being widely circuitous, should
at the expense of a few miles' extra distance save the necessity of the cuttings and tunnels which a route through
the chalk ridges, known as the North and South Downs,
would involve. The most prominent point of distinction
between Sir John Rennie's and Robert Stephenson's plans
may be seen in the one fact, that while the earthwork of
the latter, estimated on slopes averaging at one and a
half to one, was 6,000,000 cubic yards, the earthwork of
the former, estimated on slopes of less than one to one,
was 8,000,000 cubic yards. Thus, calculated with
Robert Stephenson's slopes, of one and a half to one, Sir
John Rennie's earthwork would at the least have amounted
to ten million cubic yards, or four millions of cubic yards
more than Robert Stephenson's.[*]

---

[*] 'Now the most important part   earthwork in these two several lines.
of the comparison is the amount of   Now the amount of the earthwork

In the following session of 1837, in accordance with a resolution of the House of Commons requesting the Crown ' to refer to some military engineer the statements of engineering particulars furnished by several engineers in support of the several lines of Brighton Railway now under the consideration of the House,' Captain Alderson, R.E., reported on the merits of the various proposed routes. In his report, dated June 27, 1837, the referee said —

I have the honour to state, that I have carefully read the evidence given before the committee, as well as their report, and attentively compared the several plans and sections committed to me; that I have also taken a general survey of the sites of the different lines, examining more attentively those portions where important works are proposed, and have no hesitation in stating that the line proposed by Mr. Stephenson, *considered in an engineering point of view, is preferable to either of the others.* Availing himself of the valleys of the Rivers Mole and Adur, he avoids the heavy cuttings necessarily consequent on forcing a passage through the chalk ridges known

---

upon Mr. Stephenson's line is 6,000,000; his slopes averaging at one and a half to one. Sir John Rennie's, with slopes averaging at less than one to one, are 8,000,000 of cubic yards. Now Mr. Stephenson has formed an erroneous estimate of Sir John Rennie's earthwork, it appears, because he assumed the same slopes as he himself took; and taking those of Sir John Rennie's at 8,000,000 cubic yards, to 6,000,000 on Mr. Stephenson's line, would, if Mr. Stephenson's slopes are adopted, become 10,000,000. If Mr. Stephenson adopted Sir John Rennie's slopes of course his earthwork would be very much diminished. And here arises a remark which applies to the whole of this case, and I do not

know that I need repeat it by-and-bye. Even supposing Sir John Rennie could do what he proposes to do, in the time and at the expense he proposes, I may admit my friend's case, and I say I can do the same. Instead of its costing me a million of money, it will cost me £700,000 : instead of taking me two-and-a-half to three years, it will be done in eighteen months. What you can do, I can do; and you derive no benefit from taking fallacious accounts of earthwork and slopes that cannot stand ; because you must admit, that what you can do I can do also.'— *Hon. J. C. Talbot's Speech before Committee of House of Commons, May* 17, 1836.

as the North and South Downs; and, with the exception of two short tunnels, one at Epsom and the other at Dorking, arrives at Brighton, viâ Shoreham, having only such ordinary difficulties to contend against as are necessarily consequent on undertakings of a similar nature and extent.

Having borne this emphatic testimony to the engineering excellences of Robert Stephenson's line, Captain Alderson entered on the examination of Gibbs's and Sir John Rennie's proposals, giving the latter the palm over both its rivals; its superiority to Robert Stephenson's line being discerned in the greater advantages which, the Captain judged, it offered to the sea-ports on the South-Coast. 'I therefore,' concludes Captain Alderson's report, 'adhere to the opinion already given in favour of the Direct Line.' Robert Stephenson was greatly chagrined at this decision. As the line between Brighton and London was one to which the attention of London residents was especially directed, his defeat had all the additional mortification which a crowd of spectators imparts to the overthrow of a combatant. Moreover, he sincerely felt that the conclusion was not only erroneous, but that it was arrived at by a one-sided consideration of the very points which ought to have led to a judgement in his favour. As he had done years before in the case of Messrs. Walker's and Rastrick's reports against the locomotive, Robert Stephenson again defended himself with his pen, and in a short pamphlet* —which is a model of criticism, in temper, conciseness, completeness and perspicuity—gave a clinching response to the fallacies of Captain Alderson's report.

---

* London and Brighton Railway. Mr. Robert Stephenson's Reply to Captain Alderson. London: John Weale, Architectural Library, 5 High Holborn: 1837.

Amongst the parliamentary contests of this period, in which a conspicuous part was assigned to Robert Stephenson, that of the London and Blackwall line, in 1836, deserves notice, as the proceedings before committees sitting upon that line exhibit both the elder and younger Stephenson as opponents of the locomotive system. Amongst the many railway projects of 1825, one much favoured by Sir Edward Banks and other capitalists was a proposal for an iron flange railway (similar to those tramways which a citizen of the United States recently placed in some of our wider thoroughfares) which should be laid down upon the Commercial Road for the acceleration of traffic to the West India Docks. As the rails were to be laid upon the public road, the employment of locomotives was of course no part of the project. The trustees of the Commercial Road liked the scheme, and with their approval a Bill for the construction of the tram was brought into Parliament in the session of 1827–28, but after two readings it miscarried through want of support. The defeat, however, did not deter the trustees of the road from doing something for public convenience. At their command Mr. James Walker in 1828 began to lay down the stone tramway from Severn's sugar warehouse, 200 or 300 yards below Whitechapel Church, to the gate of the West India Docks, the granite blocks of which tram remain to this day.* The road was

---

* Mr. James Walker, examined before the Committee of the House of Commons, on the London and Blackwall Railways, April 14, 1836. 'Committee.] All the way?—Yes, the Commercial Road rises towards London: the horse was doing as much work in drawing that ten tons, as if he had been drawing thirty tons on a level. 'Do you conceive that a horse's draught is equal to three hundred and sixty pounds on a level tram-road?—The horse did the work I

completed in 1829, and was in its day deemed a great success, although the rapid improvements made in the next few years in the construction of iron roads and steam carriages soon rendered it of comparatively small importance.

In 1835, the Company whose Bill had been thrown out in 1827–28 renewed their exertions, and, having grown bolder by experience, resolved on applying to Parliament for leave to make a new road, to be worked by steam-power. The proposal was no longer for an iron-tramway on the Commercial Road. George Stephenson was the engineer of this Company (the London and Blackwall), and at the outset advised a route along the south of the Commercial Road. Subsequently taking a different view, he induced the Company to adopt a line on the north of that thoroughfare ; upon which a new Company, styling

---

have stated. I was just going to add, he was a very powerful horse, and it was too much for him: it was a kind of trial of strength, and the horse did it. That is quite as much as two horses could have done. I will read the paragraph if you please. The full average work of a horse per day is one hundred and fifty pounds moved twenty miles over a pulley; that is, raised one hundred and fifty perpendicularly: the horse, therefore, was doing the work of two horses and a half. He appeared to go easily, but the exertion was too great to be continued for any considerable time, so as to form the basis of a calculation, but it was extraordinary work to draw on a turnpike road thirty tons. Upon the whole, I think the conclusion is, that if the roads were level the work of a London draught horse on a

tramway would be ten tons gross; but as the Commercial Road rises towards London, a deduction must be made from this for its gravity, the amount of which depends on the inclination of the road, and is common to all kinds of roads, of tramways, and railways; and, therefore, take all things into consideration, I am of opinion that six tons gross from the Docks to Whitechapel, and a greater weight from White-chapel to the Docks, may be considered as the proper weight for one horse on a tramway—six tons up, I think I may also say twelve tons down, on the tramway.

'On what supposition ? with its present inclination ? — Yes.

' You stated that this horse drew ten tons from the Docks to the end of the road ? — Yes.'

itself the 'Commercial Road Railway Company,' with Sir John Rennie for engineer, started up, and, adopting the relinquished southern route, entered into competition with the original projectors. The Commercial Road Railway would have started from Glass-House Yard, on the east of the Minories, and have passed on a viaduct supported by arches to its terminus near the Brunswick Wharf at Blackwall. This line Sir John Rennie proposed to work with locomotives. On the other hand, George Stephenson maintained that no line passing through such dense and valuable property as that which lay on the banks of the river between London and Blackwall ought to use steam carriages, on account of the danger from sparks, which he apprehended might cause serious losses by fire. Robert Stephenson* fully concurred with his

---

* House of Commons' Committee on the London and Blackwall Railways, May 17, 1836. Mr. Robert Stephenson examined by Mr. Alexander.

'I want to know whether you have directed your attention to their extension in towns? — Yes; the mode of having tunnels under towns.

'Has your attention been directed to the best means of avoiding inconvenience to the property in the line of railway? — Yes; and in almost every case, after the best consideration, we have adopted stationary engines.'

.   .   .   .

'Committee.] In a crowded metropolis and a great town like this, you appear to think that the stationary engines are less objectionable than the locomotives? — I think so, de-

cidedly; and that decision was not rashly come to.'

.   .   .   .   .   .

'Mr. Alexander.] Do you know of danger having arisen from locomotive engines having set fire to property of any kind? — Yes.

'On what line? — Why I have seen them set fire to the gorse on the Liverpool and Manchester line; on the Leicester and Swannington Railway, which had a farm-house burned, the premises were consumed. I must mention there that the railway was extremely near.

'Committee.] Was the engine the cause of the fire? — The immediate cause.

'Had the Company to pay the damage? — Yes, they had.

'Do you think the proposed gauze that covered the funnel would be a remedy against all danger of this father, and gave emphatic testimony as to the hazard of setting fire to towns by driving steam carriages through them.

Thus far had the position of the locomotive changed in public estimation. The Stephensons who, ten years before, led the scanty band of its supporters, could now venture to state boldly what they regarded as its serious disadvantages, and could find courage to check the anxiety of speculators to use it under all circumstances. Doubtless there was real danger of fire from passing locomotives. A recent cause, by which a railway company was compelled to pay heavy damages for certain agricultural property ignited by sparks thrown from a steam-carriage, attests that the Stephensons had foundations

---

kind? — Not a remedy against all danger, but it would materially mitigate the danger.

'You agree with Mr. Bidder in what he stated with respect to the gauze, the making it of a sufficiently fine nature would reduce the power? — The reduction of the power by the gauze has been carried to as great an extent as is desirable, or rather more so, because it has been taken to that point that if you go beyond it the effect of the engine is reduced.

'Now, even with the gauze in that state, have you seen sparks of ignited coal carried through by the force of the draught, or blast?—Yes.

'Would that be dangerous in such places as rope-walks, or timber-yards, or ship-yards, which are full of combustible materials? — Yes; it is in consequence of the property that I objected to the use of locomotives under these circumstances.

'The danger arising from locomotives is one of the elements which led you to form your opinion as to the propriety of locomotives in the neighbourhood of a town? —Yes; I consider it of the greatest importance to bring the Blackwall Railway as near to the city of London as possible; *and when I considered the immense value of the property on every side, I came to the conclusion eventually that, rather than have locomotive engines, I thought the railway would have to be abandoned.*

'*Committee.*] You think locomotive engines through that line of property, with such a trade as is carried on, are objectionable? — Yes.

'Because dangerous? — Yes, that is the sole objection—in fact, a chief one; and the other I referred to is the great space necessary for a depôt, and the enormous expenses which are requisite for its formation.'

for their fears.   Not less certain is it that their fears were
excessive.   At the present date, when the ropes and
stationary engines with which the Blackwall line was long
worked have been for years discontinued, and when
locomotives are shooting to and fro through every quarter
of London at every variety of distance above the level of
the streets, and passing through every description of
property, a person of the humblest intelligence would
smile at an assurance that London ran any risk of being
destroyed by sparks thrown out from the chimneys of
locomotives.   That the Stephensons so miscalculated on
a point relating to the locomotive, is a matter worthy of
reflection; that their error was on the side of *caution*, is
a fact that illustrates one of their principal characteristics, and points to the cause of a large part of their
success.

Notwithstanding the steady increase in the number of
subjects demanding his attention, Robert Stephenson
resolutely adhered to a rule which he had laid down on
first settling at Haverstock Hill, — a determination to
read something every day.   He was an early riser, and
always managed to get two hours of study before breakfast.   The time was short, but he used it to such good
purpose that, with the aid of scientific periodicals, he was
always well up in the recent discoveries of science.
Mathematics, chemistry, geology, and physiology, were
his favourite departments of study.   For light literature,
his active life left him scarcely any leisure ; but few
weeks passed over in which he did not find an hour to
devote to an English poet.   For the political articles of
newspapers he cared little; so that, notwithstanding his

strong political convictions,* there were few men worse informed than he usually was on the contentions and party warfare of the Houses.

In his domestic life Robert Stephenson was, with the exception of one circumstance, a happy man. He had not misjudged the character of the lady whom he married and took to his home in Newcastle, at a time when he had no prospect of speedy advancement to eminence and wealth. The young wife, who 'ruled her husband without ever seeming to rule him,' was much liked by all Robert Stephenson's friends, and contributed in no slight measure to secure his position amongst his professional brethren by the amiability and tact with which she pre-

---

* From first to last, as boy and man, Robert Stephenson was a staunch, unswerving, uncompromising Tory — not a Conservative, but a Tory. As occasion will be taken to show in a future chapter, he never allowed his opinions on public affairs to interfere with his private friend. The conclusion of the following note to his friend Frank Forster will show how generous a politician he was : —

'Engineering Department,
'Camden Town :
'Aug. 29, 1838.

'Dear Forster,—At the meeting of our Board to-day, the importance of opening on the 17th of September was so strongly urged on the score of the immense expense and inconvenience of the coaching, that I felt bound to promise them that it should be practicable on that day. I must, therefore, lean upon you again, and I do so with confidence ; but if you should find it impossible to complete both lines, you had better at once consider the propriety of putting in some points and crossings, so that our line may be passable over that portion of the Hill Morton embankment (and through Kilsby) which may, perhaps, not be quite closed. The state of the permanent road is not of so much consequence as the existence of it throughout, so that the trains may be able to pass upon it at a very slow speed. I shall be coming down to Coventry on Saturday night or Sunday morning. Could you drop me a line by return to say where I shall meet you ? Will you go to Cracow to make a line from that place to Warsaw ? I have an application ; but I fear your Liberal principles will give rise to some objection on the part of the Autocrat of all the Russias. The line is 100 miles in length. More of this when we meet: in the meantime try if you can't convert yourself into a Tory.'

sided over a household where men of incongruous
dispositions and rival interests frequently met. The cloud
over the domestic life of Mrs. Stephenson and her husband was their want of children. Robert Stephenson
greatly desired to become a father, but his wish was not
to be gratified. The part assigned to him was to conceal his disappointment from his wife, and to find cheerful companions for her in that home which was never to
be musical with the prattle of babes. To achieve this
latter purpose he encouraged her to surround herself
with the members of her own family. One of her relations became the commercial manager of the Newcastle
factory and Robert Stephenson's confidential agent in all
his North Country affairs. Another, a young lady, was
an almost constant visitor at Haverstock Hill.

A week seldom passed in which Robert Stephenson
neglected to write to Newcastle; and in his letters despatched to 'the works,' occasional glimpses are caught
of his private habits and amusements. Relating for the
most part to orders for new engines, the solvency of commercial houses, and other ordinary topics of business
interest, they occasionally contain scraps of information
relative to Mrs. Stephenson's doings and wishes, all such
passages being pervaded by a spirit of simple manly love,
and standing out all the fresher and brighter for the
prosaic communications in which they are bedded. In
one letter Robert Stephenson enquires anxiously about a
'Smuggler' to be sent him from Newcastle, the said
Smuggler being a painting which he wished to add to his
collection of art treasures, already growing numerous.
In another letter he is earnest about the qualities of a
new horse. In those days he thought £60 a rather high

price for a horse. A third long letter (dated Feb. 27, 1837) concerning boilers and prices, Robert Stephenson finished off with an additional word about the 'Smuggler,' and one curt line on his domestic affairs : ' We are all tolerably well at Hampstead ;' when the pen was snatched from his hand by the young lady before mentioned, and a post-script added—

My Dear Uncle,— Cousin Fanny would have filled up this part, but she is in bed with a sick headache. Tell Mr. Hard-castle Mr. G. Stephenson's brother Robert is dead, the new groom has been thrown from his horse, and both horse and man are at present perfectly useless. This is what Mr. Stephenson calls being tolerably well at Hampstead.

About two months later another of those prettily or-namented business letters to ' Uncle Edward ' contained a sketch and three or four lines from Mrs. Stephenson's pencil. A few days before, Mrs. Stephenson had met with an unusual accident. She was driving from a friend's door, where she had been making a call, when she stood up in her phaeton, and, looking backwards, waved and nodded another ' good-bye' to some acquaintances at the drawing-room window. Scarcely had she done this when she fell back on the seat of her carriage, frightened and faint, and saying she had broken her knee. On examina-tion it was found that the ligament uniting her right knee-cap to the muscles of the thigh had given way During her tedious cure Mrs. Stephenson had to lie night and day on a double-incline bed, and in the rather awkward posture which that couch compels, she drew a humorous picture of herself.

The men with whom Robert Stephenson was most familiar at this period were his firm friends throughout

life. Amongst them were Mr. Bidder, Mr. Thomas
Longridge Gooch, Mr. Budden (who acted as his secretary), Mr. John Joseph Bramah, Mr. Frank Forster, Mr.
Birkinshaw, and Mr. Charles Parker. Robert Stephenson
was a man of few pleasures. Music he cultivated to a
certain point to please his wife; but at this period he
rarely touched his flute. His profession was to him both
business and pleasure. On Sundays, however, he relaxed.
In the morning he usually went to church. In the afternoon he wrote letters and took a walk, and finished up
the day with receiving a few professional friends at dinner, immediately after which the cigar-box made its appearance.

Amongst Robert Stephenson's more distinguished associates at this period was Professor Wheatstone, the joint-
inventor with Mr. William Cooke of the electric telegraph. Their memorable invention was patented in June
1837, and before the autumn of that year was at an end,
the correspondence necessary for business purposes between the Euston Square and Camden Town stations was
carried on by electricity. 'Robert Stephenson's London
and Birmingham line' has the honour of being the scene
of the first successful working of electric telegraphy.

In Dr. Andrew Wynter's 'Curiosities of Civilisation,'
the following interesting passage occurs in the article on
the 'Electric Telegraph':—

Following up his experiment, Professor Wheatstone worked
out the arrangements of his telegraph, and having associated himself in 1837 with Mr. Cooke, who had previously devoted much
time to the same subject, a patent was taken out in the June of
that year in their joint names. Their telegraph had five wires
and five needles; the latter being worked on the face of a lozenge-
shaped dial, inscribed with the letters of the alphabet, any one

of which could be indicated by the convergence of the needles. This very ingenious instrument could be manipulated by any person who knew how to read, and did not labour under the disadvantage of working by a code which required time to be understood. Immediately upon the taking out of the patent, the directors of the North Western Railway sanctioned the laying down of the wires between the Euston Square and Camden Town stations, and towards the end of July the telegraph was ready to work.

Late in the evening of the 25th of that month, in a dingy little room near the booking-office at Euston Square, by the light of a flaring dip-candle, which only illuminated the surrounding darkness, sat the inventor, with a beating pulse, and a heart full of hope. In an equally small room at the Camden Town station, where the wires terminated, sat Mr. Cooke, his co-patentee, and, amongst others, two witnesses well known to fame—Mr. Charles Fox and Mr. Stephenson. . . . . Mr. Cooke in his turn touched the keys and returned the answer. ' Never did I feel such a tumultuous sensation before,' said the Professor, ' as when all alone in the still room I heard the needles click; and as I spelled the words, I felt all the magnitude of the invention, now proved to be practical beyond cavil or dispute.' The telegraph thenceforward, as far as its mechanism was concerned, went on without a check, and the modifications of the instrument, which is still in use, have been made for the purpose of rendering it more economical in its construction and working, two wires at present being employed, and in some cases only one.

Professor Wheatstone, whilst making valuable communications for the purposes of this work, bore emphatic testimony to the zeal displayed by Robert Stephenson from first to last—from 1837 up to the time of his death — to advance the science and protect the interests of telegraphy.*

---

* Those who are curious in the history of the telegraph will find a distinct proposition for a system of telegraphic inter-communication of thought in the ' Scot's Magazine' (vol. xv. p. 73) of February 1753.

Since he fixed himself in town Robert Stephenson had enjoyed a fine and rapidly increasing professional income —an income to be measured by thousands. He had, therefore, begun to live with the luxury and some of the ostentation, usual with persons of wealth In compliance with Mrs. Stephenson's wishes, but not without reluctance, he visited the Heralds' College, and informing the heralds that, according to a family tradition, he was descended from ' the Stephensons of Mount Grenan in Scotland,' asked permission to use the arms of that house. In what estimation the officers of the college held ' the tradition ' it is needless to enquire. On the whole, they acted with discretion. Taking a middle course between their own interests and the rights of the Mount Grenan Stephensons, they took some of the fleur-de-lis and mullets from the shield of the Mount Grenan family, and, having dished them up with a crest and other garnishings, granted them as an heraldic bearing to Robert Stephenson and his father and their descendants. These arms Robert Stephenson took (November 21, 1838), and without haggling paid the sum at which they were priced. Honestly bought, they were perhaps obtained not less honourably than many ancient devices tricked in the College archives. But Robert Stephenson, truthful, honest, and simple, with a repugnance to flattery and a detestation of shams, never liked them.

Not long before his death, his eye chancing to fall on an object ornamented with his arms, he blushed slightly, and said to an old friend by his side — ' Ah, I wish I had n't adopted that foolish coat of arms ! Considering what a little matter it is, you could scarcely believe how often I have been annoyed by " *that silly picture.*"'

CHAPTER XII.

FROM THE COMPLETION OF THE LONDON AND BIRMINGHAM
RAILWAY TO THE OPENING OF THE NEWCASTLE AND DAR-
LINGTON LINE.

(ÆTAT. 35–41.)

Railways undertaken in various Directions — Brunel, Giles, Braith-
waite—Robert Stephenson's Trip to Italy—On his Return again
immersed in Projects — The Contractors' Dinner at 'The Albion'—
Letters to Newcastle — Cigars for the Continent — Stanhope and
Tyne Crisis — Robert Stephenson threatened with Insolvency—Acts
for the Pontop and South Shields and the Newcastle and Darlington
Junction Railways — Robert Stephenson appointed to execute the
Newcastle and Darlington Lines — Robert Stephenson created a
Knight of the Order of Leopold — Mrs. Stephenson's Death —
Opening of Newcastle and Darlington Line — Public Dinner and
Speeches — Continental Engagements—Leaves Haverstock Hill and
moves to Cambridge Square — Fire in Cambridge Square — George
Hudson and Robert Stephenson — A Contrast.

THE railway system was fixed. To disturb that system
attempts were made by men of intellect and high
character ; but those attempts were futile. The principal
rules laid down by the Stephensons between 1820 and
1838 are the rules of railway engineering at the present
day. The example set by the great leaders was followed
successfully in all directions. The younger Brunel, a man
dear to all lovers of genius, was at work on the *Great
Western* ; Mr. Francis Giles was laying down the line
between London and Southampton ; Mr. John Braithwaite
undertook the *London and Colchester*, bringing life and

increased trade to the eastern counties. In the north, George Stephenson had under his supervision the *Manchester and Leeds*, the North Midland from *Derby to Leeds*, the *York and North Midland* from Normanton to York; the Grand Junction Railway projected by the father (but executed by Joseph Locke) having already united his magnificent Liverpool and Manchester line with his son's road terminating in Euston Square. It would be beside the purpose of this work to enter into the details of each of these works, and of the other lines that followed them in quick succession—details for the most part closely resembling each other. It will be sufficient to select for description those roads alone on which Robert Stephenson's distinctive powers found most emphatic expression.

The engineer-in-chief's labour on the London and Birmingham Railway was by no means at an end when the line was opened. Works on it still remained to be completed, and improvements had to be made at various points before ' the chief' (as up to the day of his death Robert Stephenson's staff were wont to call him) could dismiss the line from his thoughts. As soon as he was able to give his attention to the matter, the North Midland line from Derby to Leeds was on his hands. He was also needed on the continent. The grand cross lines from Ostend to Liege and from Antwerp to Mons were under construction and requiring his superintendence. Invitations also reached him to visit France, Switzerland, and Italy, to advise on lines contemplated in those countries. Entrusting the superintendence of his home lines to his father and the execution of them to his subordinates, and quitting Westminster when the business of the committee-rooms was daily becoming heavier, he left England for three months to answer in person these calls from foreign countries. At this period he became intimately acquainted with Mons. Paulin Talabot, a civil engineer who for many years has held a leading position amongst the civil engineers and capitalists of the continent. At a subsequent period Robert Stephenson, Signor Negretti, and M. Talabot surveyed the Isthmus of Suez, and ascertained that the levels of the Mediterranean Sea and the Red Sea were identical.

On his return he was soon busy again with the affairs of English railways. On July 1 he attended the meeting of the Council of the Railway Society. The next day saw him giving evidence before the Select Committee of Railways. On the 16th of the same month he was at Derby about the railway station of that town; the next day at Clay Cross to look over his father's coal mines; the next day at Sheffield and over the *Sheffield and Rotherham* Works; the next day at Tapton to negotiate the purchase of land for a railway station; the next day at Birmingham to meet the Committee of the London and Birmingham line; the next day in town for examination before parliamentary committees.

In the autumn of this year, Robert Stephenson received an expression of the high esteem in which he was held by an influential division of the business men engaged in the construction of the railways of the country. As the reader is by this time well aware, a large number of the contracts on the London and Birmingham line came back to the Company uncompleted. Of course the contractors did not get quit of their engagements without much delicate and painful negotiation with the directors. In other cases where the contracts were fulfilled, the course of their performance was marked by misunderstandings and disputes between the Company and the master-employers. To arbitrate in such disputes, and to adjudicate in such difficulties, Robert Stephenson was by temper, information, and reputation, peculiarly fitted; and it adds not a little to his fame that in nearly all the disagreements between directors and contractors he was appointed sole umpire.

The course thus commenced on the London and Birmingham line was continued on the North Midland, the Derby Junction Railway, and the York and North Midland. Whenever a contractor on one of his lines was contending with directors about the terms of an agreement, it was left with Robert Stephenson to arrange the difference.

Such services merited signal reward; and in 1839 a movement was set on foot to make an appropriate acknowledgement of them. A party of gentlemen, who were assembled (April 2, 1839) in Birmingham on a different business, suggested the propriety of presenting the popular engineer with a testimonial. The proposition was so well received, that before the meeting separated the affair had been well started. A committee, with Mr. J. D. Barry, of Manchester, for honorary secretary, had been appointed, with powers to ask for subscriptions — it being arranged that no contribution should exceed £5 and that no one should subscribe but 'gentlemen who had been engaged as contractors for the construction of railways or for the supply of permanent materials. A sum of £200 was subscribed in the room, and by the following November the committee held more than £1,250 for the accomplishment of their object.

In the previous July a committee of taste had been appointed to decide on the form of the testimonial. In this committee Sir John Guest, M.P., and Mr. Crawshay represented the iron trade, Mr. Bramah and Mr. Maudslay the engine manufacturers, Mr. Freeman and Mr. Bazley White the stone and cement trades, Mr. Dowson and Mr. Holland the timber trade, Mr. David M'Intosh and Mr. Thomas Jackson the operative railway contractors. They selected a service of plate, of which the principal ornament was a candelabrum.

This service was presented to Robert Stephenson on Saturday, November 16, 1839, when he was entertained at a grand dinner in the Albion Hotel, Aldersgate Street. The banquet was attended by more than two hundred gentlemen, several of whom came from Lancashire. In the absence of Mr. Crawshay, who was to have taken the chair, Mr. William Routh (Mr. Crawshay's partner) presided, the vice-presidents being Messrs. Joseph Dowson, John Joseph Bramah, Thomas Grissell, and Thomas Jackson. On the chairman's right sate Robert Stephenson, the hero of the evening. On the left of the chairman was George Stephenson. At present, the father had received no similar acknowledgement of his services, as ' the Author of the Railway System.' Indeed, his achievements had for the moment been eclipsed by those of his son. The fine old man, whose kindest teacher had been adversity, was even yet not duly appreciated in the metropolis. His manners were rugged and far from prepossessing, and his personal connections were for the most part in his own ' old country.' For one inhabitant of London who visited the Liverpool and Manchester line, ninety and nine were familiar with the works on the London and

Birmingham Railway. Moreover, the father, with the appearance and reputation of having seen more years than he actually numbered, was in the decline of life, whilst the power and fortunes of the son were in the ascendant. It is therefore easy to account for the fact, that 'the Father of the Railway System' saw his son thus publicly honoured, whilst he himself had been comparatively un-noticed.

George Stephenson had still to wait for his 'railway testimonial;' but not the less was he delighted with his son's triumph. Indeed, why should he grudge 'the lad' the world's homage? To make him 'a great man' had been his aim from the time when he wrote down the boy's name as 'engineer' on the plans of the Stockton and Darlington line.

In his life of turmoil and many cares Robert Stephenson had few opportunities for domestic repose. Whenever he could manage to do so, he spent Sunday at Haverstock Hill, often posting for hours in order that he might have a quiet day with Mrs. Stephenson. From 1838 to 1844 the urgent calls upon his time in different parts of the country made his presence in his own house almost a surprise to its inmates; but whenever he was there the home was the merrier. Amongst the thousands of letters perused for this memoir, the following was found tied up with epistles relating to improvements in the locomotive and the execution of orders at the Newcastle factory.

Feb. 1840.

DEAR EDWARD, — I bought when last in Newcastle two plaids, which have been intensely admired, and this compels me to venture on troubling you to purchase two more of the same pattern. The ladies have determined upon sporting plaids of this character in the precincts of the metropolis, and in a season or two they expect to be designated Scotch lassies. The above sketch is made with the view of guiding you in selecting the same pattern, by which you will perceive that the ground is green with stripes of red and what I call black, but what I call the black stripes seems to partake of the qualities of the chameleon, for Fanny declares it to be lavender. Now I am obliged to confess total ignorance of this peculiar colour lavender; suffice it to say I still consider it to ordinary eyes black. I would therefore advise you, if you meet with a colour between black and lavender, to consider that you have hit the mark. I purchased these said admired plaids at Robson and Henderson's, and they may perhaps recollect a strange outlandish-looking gent purchasing two plaids, and requesting them to be forwarded to the Queen's Head. I mention this as a sort of collateral aid to you in your commission, which I fear you will consider a difficult one. The sketch shows the exact distances and the relative widths of the stripes, for it is made by laying the plaid upon the sheet of paper.

Should you succeed, you will be good enough to send them by coach. I paid one sovereign each for mine, and they are of fine quality. This is essential, as they are infinitely warmer than coarse ones.

Fanny, I think, is going on well, although she is still grazing on macaroni, and occasionally a little marine flesh, vulgarly called fish. She desires to be kindly remembered to all, and requests me to say all the —— are well.

Yours sincerely,

Rob. Stephenson.

E. J. Cook, Esq., Newcastle-upon-Tyne.

By this time the incumbrances of the Stanhope and Tyne Railway had grown with fearful rapidity. In 1839 the portion of line between Stanhope and Medomsley was declared to be so ruinous, that the directors determined no longer to work it. The railway being thus disused, the lime quarries held on lease at a rent of £2,000 per annum could no longer be worked with

even a semblance of profit, and were consequently permitted to lie idle. Thus nearly the whole of the
original scheme was deserted as bad, when its desertion
entailed on the speculation an annual payment of rent
amounting to £2,300. By the end of the following year
(1840) the debts of the Company were so great that
creditors began to clamour. Bills for which the Company was responsible were floating about in all directions,
and the holders refused to renew them.

Until the close of 1840, Robert Stephenson was in
ignorance of the exact nature of his position. He knew
that, as payment for professional services, he held shares in
a line that had turned out badly; and somewhat vaguely
he had for two years feared that the consequence of his
holding £1,000 of stock would be that he would have to
pay up some hundreds more to make an arrangement
with the creditors of the association. In 1840, however,
he learnt to his horror that, as a shareholder, he was
personally responsible for the entire debts of the undertaking. It was also frankly intimated to him that, unless
certain bills were met on their falling due, the holders
would come upon him for the money. He could
scarcely credit the announcement. Without his knowledge insolvency had long been staring him in the face.

Of all the shareholders, he was perhaps the one from
whom the creditors of the Company were most secure of
payment. It was known that he had for years been
earning a magnificent income, not one half of which he had
expended on his pleasures or his establishment. It was
known also that he disliked speculation, and invariably
put his savings into investments that were secure, and
could be easily realised. His professional position and

public reputation also would spur him to a prompt liquidation of legal claims on his purse. To him, therefore, was brought the first bill which the directors could not meet. To Robert Stephenson, who throughout life was strangely ignorant of the simplest rules of law, the application at first seemed little more than an awkward joke. He soon regarded the affair in a different light; and coming in great agitation to his friend and solicitor Mr. Parker, asked what ought to be done. Writing to Mr. Cook at Newcastle on December 2, 1840, when the blow was still new, Robert Stephenson said—

> I hope you will be able to make a dividend soon : I wish this for two reasons — firstly, because I want money, and secondly because I don't like your bankers.* If they are not speculating beyond what is prudent I am deceived. And in that opinion I am borne out by several circumstances which have lately been brought before me in a way likely to affect myself very seriously. That prince of rogues has, I am sorry to say, involved all parties connected with the Stanhope and Tyne and almost all the banks of Newcastle and Sunderland. When I first became acquainted with the awful responsibilities which the Stanhope and Tyne had incurred, and the utter inability of the concern to meet them, I was perfectly stunned, and your bank has lent them on bills £51,000, which are at this moment floating. Some become due next Saturday, on which day, I have no doubt, the Stanhope and Tyne Company must stop payment. This is exactly what I am most anxious to do, for to allow —— and —— to proceed further would be madness. I have got parties in London to write down to the bankers not to accept or renew any more bills. —— left town last night for the purpose of getting them to do so, but I expect I am in time to stop his reckless career.

At Mr. Parker's advice, an extraordinary general meeting of the proprietors was summoned for December

---

* This suspicion, as the business men of Northumbria can testify, was signally justified by subsequent occurrences.

29, 1840, which was adjourned to January 2 next following. At this adjourned meeting it was decided to dissolve the Company, and to form a new Company with a capital of £400,000, such new concern taking, at the same time, the property and the debts of the bankrupt association and applying to Parliament for incorporation. This bold plan offered the shareholders the only chance of extrication from their embarrassments. The intention of the new projectors was to apply their subscribed capital to the immediate liquidation of their debts, and by stringent economy to endeavour to carry on the concern without loss. If they should succeed, all would be well. But even if failure should be the fate of the new Company, acting under parliamentary sanction, individual shareholders would be defended from ruin.

There was no room for loss of time. It was necessary that the capital of the new association should be subscribed and paid up without delay, for creditors were importunate. On February 5 another meeting of shareholders was held, when the resolutions of January 2 were confirmed and the old Company was dissolved. At the time of the dissolution, there were forty-nine interests in the proprietary of the Stanhope and Tyne Company. Of these thirty-six absolutely consented to the dissolution; of five other interests, where the original holders were dead, four executors gave in their consent, the fifth executor being abroad and not opposing. Four other shareholders were bankrupt.

Of course all the shareholders were invited and urgently pressed to subscribe to the new speculation. Equally a matter of course was it, that the smaller shareholders were disinclined to embark their hundreds on another

venture between Stanhope and Tyne. They were only
too glad to put their liabilities on shoulders stronger than
their own.   It remained for the greater men with greater
interests at stake to advance their thousands on the effort
of retrieval.   The assets of the dissolved Company, at a
liberal computation, did not exceed £307,383, whilst the
liabilities of the affair were £440,852.   To deal with this
accumulation of debt, the monied men made great efforts
to contribute effectually to the capital of the new undertaking.

On the committee of the new Company Robert
Stephenson's name was placed, and, regarding the venture
as the only possible means of escaping from his perilous
position, he threw himself heartily into its interests.   The
line of action once decided on, his mind became easier ;
and with the soothing assurance that the best measures
had been adopted, he resolved to persevere in them and
await their result with calmness.   On the evening of the
29th he wrote to Newcastle.

35½ Great George Street, Westminster : Dec. 29, 1840.

DEAR EDWARD,— We have this day had a meeting of the
Stanhope and Tyne Railway, and I hope its result portends
good, but it is still somewhat uncertain. The Company's affairs
are awfully deranged, and the precise consequences no one can
venture to predict. We may possibly struggle through, but this
hope may prove fallacious. . . . . I dare say I shall very
shortly be at Newcastle, when I can explain more at length all
the outs and ins in this affair, which are too painful and too
lengthy for an epistle of an ordinary character. When I was
last at your canny town I intended to have gone into the
matter with my friend Stanton, * but, as you saw, I was too

---

* Mr. Philip Stanton—one of the many friends whom Robert Stephenson
attached to himself in early life, and kept close to his heart till death.

much occupied and too anxious to sit down to talk over matters involving such consequences. The history of the Stanhope and Tyne is most instructive, and *one miss* of this kind ought to be, as it shall be, a lesson deeply stamped. If the matter get through, I promise you I shall never be similarly placed again. Ordinary rascality bears no relation to that which has been brought into play in this affair. I conclude from what has transpired that all the —— will shortly be in the Gazette, and how many they may drag after them into the same position it is impossible to predict.

On January 12, just ten days after the adjourned general meeting, £250,000 of the proposed capital of £400,000 were subscribed for the new association, the rest of the required sum being in due course found. Robert Stephenson put his name down for £20,000 and, like his fellow-subscribers, promptly paid the amount of his subscription as the instalments agreed upon became due. To fulfill this engagement he had to raise money by all the means in his power, and to transfer one half of his share in the Newcastle factory to his father.

Hampstead : Jan. 4, 1841.

DEAR EDWARD,— Your view as to my wishes respecting one half of my interest in the factory is exactly what I wish. The transaction is not intended to be otherwise than *bonâ fide* between my father and myself. The fact is, I owe him nearly £4,000, and I have not now the means of paying him as I expected I should have a month or two ago. All my available means must now be applied to the Stanhope and Tyne. On the 15th of this month I have £5,000 to pay into their coffers. The swamping of all my labours for years past does not now press heavily on my mind. It did so for a few days, but I feel now master of myself; and though I may become poor in purse, I shall still have a treasure of satisfaction amongst friends who have been friends in my prosperity. The worst feature in the case is the all-absorbing character of my attention to the rectification of its embarrassments, which if produced by legitimate misfortune

would have been tolerable, but when produced by . . . . men
who are indebted to me, they become doubly afflicting. I am
not without hopes that before the 15th of this month we shall
have succeeded in bringing the affairs into a ta..gible state, and
about that date I hope to be in Newcastle.

Yours sincerely,
Rob. Stephenson.

Anxious to get rid of a bad name as well as bad fortune,
the new association petitioned Parliament for incorporation
under the title of 'The Pontop and South Shields Railway
Company.' Their prayer was successful, but they did not
gain the Royal assent without a struggle. The same evil
influence which had brought the affairs of the Stanhope
and Tyne into so disastrous a condition, opposed to the
utmost the Pontop and South Shields line. Through
that influence, a petition was concocted imploring
Parliament that the Act desired by the new Company
should not become law. Readers unacquainted with the
daring of reckless and embarrassed speculators, and
ignorant of the ease with which such persons can work
on the passions of the ill-informed, will scarcely believe
that a most vigorous opposition was maintained against
so honest and necessary a project. The opponents of
course represented that the 'Pontop and South Shields'
scheme was simply a conspiracy on the part of the rich
to oust the poor shareholders from an undertaking just
as it was about to become profitable. Absurd as such a
charge was, it gained so much credit that on the second
reading before 'the Lords,' Lord Canterbury denounced
the Bill as 'a measure of spoliation.'

Amongst other plans for effecting their objects, the
directors of the 'Pontop and South Shields' line entered

into a compact with the projectors of the Newcastle and
Darlington Railway, which line had for some time been
sketched out by those who were bent on uniting the
Tyne and the Thames by an iron road.   The chief points
of the contract can be stated in a few words.   Five miles
of the Stanhope and Tyne line formed a connecting link
between the Durham Junction Railway and the Brandling
Junction Railway, and were used for conveying passengers
between the towns of Newcastle, Gateshead, South Shields,
Sunderland, and other important places.   It was proposed
to make these five miles of railway a part of the Great
North of England Railway, which was to unite Newcastle
with London and the southern and western parts of the
country.   Of course such a proposal took from Robert
Stephenson and his brother directors a heavy weight of
anxiety, opening up to them, as it did, a profitable market
for a portion of their encumbered property, and ensuring
them powerful cooperation in their approaching parliamentary battle.   The two Companies agreed to assist
each other.

This arrangement is worthy of notice ; for George
Hudson was appointed chairman of the Newcastle and
Darlington Junction line, and to watch the Bill through
Parliament in the session 1841--42 he quitted York and
came up to London, where he speedily became powerful
in the railway world.   One of the consequences of the
arrangement between the two Companies was that Robert
Stephenson became the engineer of the line in which the
Railway King was interested.

It was an anxious session for Robert Stephenson, and
he had reason to dread the advent of George Hudson
upon the scene.   Fortunately, however, the chairman of

the Newcastle and Darlington Junction Line was the fit
man for the occasion.   Human nature round Westminster Abbey closely resembles human nature round
York Minster ; and when Hudson entered the committee-
rooms of the House of Commons, he found the weak,
there as elsewhere, obedient to the stronger will.   He
found throngs of men, eager, grasping, shrewd, and unscrupulous, but lacking the nerve and definite purpose
which are necessary for success in commercial gambling ;—
he saw, in fact, an army without a commander, yet sorely
wanting one.   Delicacy and fine tact would have stood in
Hudson's way.   The leading characteristics requisite for
a chief over such men, battling and struggling on a new
field of enterprise without organisation, are imperious
temper, shrewd selfishness, and fierce bull-dog resolution
to overcome every antagonist.   These qualities George
Hudson possessed in an eminent degree, as well as a
natural force of intellect beyond that of the inferior sort
of ordinary men.

The result of the session was fortunate alike for Robert
Stephenson and George Hudson.   The Pontop and South
Shields and the Newcastle and Darlington Junction Bills
both passed; and Robert Stephenson not only saw his
way to the end of the Stanhope and Tyne difficulty, but
secured the direction of a work destined to afford him
much gratification.   To connect London and Newcastle
by railway communication had long been a favourite
object of Robert Stephenson's ambition, and now he was
called upon to construct the last link of the chain.   Con-
gratulations poured in upon him from all sides.   'Many
thanks,' he wrote to one friend, ' on your kind con-
gratulations on the Stanhope and Tyne business; for

next (if not equal) to one's own pleasure in one's own
success is that of knowing our friends participate in it.'

Before the Stanhope and Tyne affair is dismissed from
consideration, it may be well to state that the Pontop and
South Shields line turned out a great success, and was, in
the course of a few years, sold on good terms to the
Newcastle and Darlington Junction Railway Company.

The anxious and trying year of 1841 was marked
to Robert Stephenson by one pleasing event. In the
August of that year the King of the Belgians created him
a Knight of the Order of Leopold, the honour being
conferred as 'a testimony of his Majesty's satisfaction
with improvements made in locomotive engines, which
improvements have turned to the advantage of Belgian
iron roads.'

The cloud of the Stanhope and Tyne trouble had,
however, scarcely been dispelled, when a far darker cloud
took its place, and Robert Stephenson was called upon
to endure the great sorrow of his life. Mrs. Stephenson
had for two years suffered from malignant cancer, when
she expired, without experiencing the weary duration of
agony which that malady sometimes inflicts upon its
victims. Tender and true to the last, she studied to
lighten the blow which was soon to fall upon Robert
Stephenson, and which, as he long afterwards remarked
to a friend, took away from him 'half his power of
enjoying success.' When he was created a Knight of the
Order of Leopold she, already too sick to care for earthly
honours, feigned the pleasure she would, a few years
earlier, have really felt at the distinction. When the
session of 1841–42 secured him from threatened insolvency, and commissioned him to lead the Northern Railway

into his 'dear old canny town,' no one exulted more at his triumph than the gentle woman who knew full well that, while the sods were being cut to make way for the new road, the turf would be raised for her own grave.

Even while such a calamity was impending, Robert Stephenson could not defer the claims of business. His note-book and letters during the summer months of 1842 show him passing from place to place along the route of the Darlington and Newcastle road, posting from one midland town to another to be present at important negotiations, and, when he was in London, working more than twelve hours out of every twenty-four over calculations, plans, estimates, and the burdensome correspondence entailed upon him by his many engagements, from Wales to Hull, and from Northumberland to the South of Europe.

On September 17 he returned to London from Cardiff, where he had been for two days examining the docks, and immediately on reaching Great George Street had a consultation about the Hull Docks. The 18th (Sunday) was spent alone with Fanny: but the next six days were devoted to business. These are the engagements and objects of attention jotted down in his note-book.

19*th*. — Hull Docks — Darlington and Stockton Bridge — French Railway.

20*th*. — French Railway — Bute Docks — Darlington Bridge.

21*st*. — French Railway Report.

22*nd*. — French Railway Report — Hull Docks.

23*rd*. — French Railway Report — Hull Docks.

24*th*. — French Railway Report — Hull Docks.

Sunday (the 25th) was, like the preceding day of rest, spent with Mrs. Stephenson; but during the three suc-

ceeding days the French Railway, the Stockton Bridge,
and the Pontop and South Shields line, occupied most of
his time and energy. The spaces allotted in the diary to
the next five days are filled up with 'At home—Fanny
very ill.' And then at the date of October 4, standing
out in affecting contrast to the brief memorials of
enterprise and labour by which it is surrounded, is the
following entry in Robert Stephenson's hand:—'My dear
Fanny died this morning at five o'clock. God grant that
I may close my life as she has done, in the true faith, and
in charity with all men. Her last moments were perfect
calmness.' On the following Tuesday (October 11) she
was interred in Hampstead churchyard, where in after
years her husband often came to stand alone and indulge
in solemn meditation. She wished him to marry again,
and on her death-bed urged him to do so. It was the
only wish of hers with which he did not comply.

Another extract from the note-book will show how he
was literally dragged from his wife's grave to the turmoil
and agitation of business.

Oct. 11*th.*— Funeral of my beloved wife.

12*th.* — Home — Stockton and Darlington Bridge.

13*th.* — Stockton Bridge, plans and specifications — West
London, estimate and plans — French Railway Report with
Berkley — Maidstone Bridge.

14*th.*— Maidstone Branch with Bidder — Norwich plans —
Newcastle and Darlington plans — French Railway Report.

15*th.* — Maidstone Branch with Bidder, and returned to
London.

During the next two years he had perhaps more work
on his hands than at any other time of his life; but of all
his engagements—the continental lines, the docks, and
his home railways—the task just then nearest to his heart

was the construction of the Newcastle and Darlington Junction, the line which would unite the metropolis with his 'ain countree.'

Two years saw the necessary works for effecting the junction begun and ended, and on June 18, 1844, the line was opened with general rejoicing and a public reception of the two Stephensons at Newcastle. The population on the banks of Tyne displayed great excitement. Bells were rung, cannon fired, and triumphal arches raised. Processions of workmen headed by bands of music, marched up and down the precipitous streets of the two boroughs, which throughout the day were crowded by the inhabitants of the surrounding colliery villages. The London 'Morning Herald' came to Tyneside that day within eight hours after its publication—a feat never before achieved. Antiquarians, who abound in Newcastle, ferreted up old newspapers, letters, and account-books, throwing light on the means of transit enjoyed by their ancestors. Copies of the following advertisement (inserted in the 'Newcastle Courant' in 1712) were, with many other interesting scraps, handed about, and in due course enlivened the columns of the local papers :—

Edinbro', Berwick, Newcastle, Durham, and London stage-coach begins on Monday, October 13, 1712. All that desire to pass from Edinbro' to London, or any place on the road, let them repair to Mr. John Baillie's at the Coach and Horses at the Head of Canongate, Edinbro', every other Saturday, or to the Black Swan in Holborn every other Monday, at both of which places they may be received in the stage-coach, which performs the whole journey in *thirteen* days without any stoppages (if God permit), having eighty able horses to perform the whole journey, each passenger paying *four pounds ten shillings*, allowing each passenger 20 lbs. of luggage ; all above, sixpence per lb. The coach sets off at six o'clock in the morning.

The Darlington and Newcastle line is by no means devoid of engineering interest, for one of its principal works is the Victoria Bridge, which spans the river Wear and the rich valley watered by that important river. Built of stone, this beautiful bridge will probably exist a memorial of Robert Stephenson's capacity, when his later viaducts of more stupendous dimensions, and carried through the air at far greater heights, shall, in consequence of their less durable material, live only in history. A fairer monument no engineer could desire. Surrounded by scenery of uncommon loveliness, its bright arches thrown from ridge to ridge (one of them leaping at a bound the entire width of the navigable river, the others spanning the fat pastures and wooded ascents on either side of the valley), present a spectacle singularly expressive of the grace and power of genius. No excursionist to the North of England should fail to leave the train at Washington, and spend a few hours at the base of Pensher Hill, in the valley of the Wear. On the summit of the hill is the ill-designed monument to the memory of Lord Durham, whilst rising from the ground beneath are Robert Stephenson's elegant curves of massive stone.

An important part of the celebration at Newcastle on June 18, 1844, was the dinner in the Town Hall, of which about 350 gentlemen partook. George Hudson was in the chair, Mr. Davies, the vice-chairman of the Company, officiating as vice-president, and Mr. John Bright, the present member for Birmingham, being one of many notable persons present.

The speech of the day was made by George Stephenson. The Hon. H. T. Liddell, M.P., on proposing the health of

the father of 'the railway system' gave utterance to the
following erroneous statement.

Of all men now living, concluded Mr. Liddell, none have
conferred so great an amount of practical benefit on society as
his respected friend, and their admired guest, Mr. Stephenson,
who, aided by strong natural talents, commencing from a working engineer to a colliery in this neighbourhood, constructed the
first locomotive that ever went by its own spontaneous movement
along iron rails.   (Applause.)

To appreciate the full force of this blunder the reader
must bear in mind that the Liddells reside in the immediate neighbourhood of Newcastle; that Sir Thomas
Liddell (afterwards Lord Ravensworth), as one of the
'grand allies,' was amongst George Stephenson's early
employers; and that it was on the property of the 'grand
allies' that the locomotive (to which the speaker referred)
was used, after it had been built with Sir T. Liddell's
money subsequent to the use of locomotives running with
smooth wheels on smooth rails at Wylam.   When a man
of high character and ability could be so misinformed not
only as to the history of the locomotive, but as to facts
that occurred almost within gunshot of his father's park,
readers need not wonder at the prevalence of the popular error which attributes to George Stephenson the
invention of the locomotive.

Anxious that his father should be the principal hero of
the day, Robert Stephenson, on his health being drunk
with a tumult of applause, spoke no more than a few
sentences, observing in the course of his brief reply that
'It was only ten years since he left the North to execute
the London and Birmingham Railway, since which time
he and his father had had the honour of being more or

less connected with every railway between Birmingham and Newcastle.'

Whilst the Newcastle and Darlington Railway was in course of execution Robert Stephenson made two visits to the continent.   In 1843 he spent several days at Naples considering railway projects, and more especially protecting the interests of the Newcastle factory from the unscrupulous competition of persons whom he had uniformly treated with liberality.   On his return home he visited various parts of Germany, securing, as his letters to Newcastle testify, new and powerful connections wherever he went.

By this time he had given up his establishment on Haverstock Hill, and moved to Cambridge Square, Hyde Park.   After Mrs. Stephenson's death he conceived a dislike for the home which he had inhabited for eight of the happiest years of his life.   It was too far from town, now that it was no longer presided over by a wife.   A widower, like a bachelor, finds it best to dwell near the clubs, so that he can readily find society.   Connected with Robert Stephenson's residence in Cambridge Square was a trifling incident, which should be mentioned, as it serves to show how careless he was about arrangements that were not connected with his profession.

Scarcely had his furniture been shifted from Haverstock Hill to Cambridge Square when much of it was destroyed by a fire that broke out in the middle of the night.   Robert Stephenson, who had only slept once or twice before in his new residence, narrowly escaped with his life from the flames.   While the house was undergoing restoration—a work that occupied nearly a twelvemonth—he took up his quarters in furnished lodgings, and had almost reconciled himself to the destruction of his property by fire, when he was greatly surprised by a demand from his landlord for the rent of the dwelling which had for ten months been unfit for use. He was not aware that in case of fire the tenant, unless he be protected by a special clause in his lease, or by the terms of a fire-insurance policy, endures the consequences of the casualty to the extent of paying rent for an unserviceable tenement.

On the night of this fire George Stephenson was sleeping in his son's house. The first in the house to sniff the smell of fire, he lost no time in taking care of himself. When Robert Stephenson and his servants were in the act of flying from the burning house in their night-clothes, the prudent father made his appearance in the hall, dressed even to his white neckcloth, and with his carpet-bag packed and swinging in his hand. This anecdote is told by friends as a story highly characteristic of his presence of mind and readiness of action.

The year 1844 is a conspicuous landmark in the career of Robert Stephenson. For twenty years he had been at work without intermission, and as the result of his exertions he found himself, whilst he was still only forty years of age, in the first rank of his profession. Had he however died then, he would have left nothing to which history could point as the monument of original and distinctive genius. He had raised the locomotive by a series of beautiful improvements from the ill-proportioned and ineffective machine of 1828 almost to its present perfection of mechanism. He had, in conjunction with his father, so fixed the English railway system in continental countries, that throughout Europe his name was identi-

fied with the new means of locomotion. His engineering
achievements were beyond all cavil works of great
ability—but not of distinctive genius.   Hitherto he had,
in the manner of a master, carried out the principles and
developed the conceptions of previous teachers, of whom
his father was the most important.   The time, however,
was now come for him to take a higher position and
accomplish works altogether without precedent.

The next six years of Robert Stephenson's life—years
memorable in the annals of social folly, crime, and
suffering—witnessed the exertions by which his influence
and name will reach future generations.   They saw the
atmospheric contest, the battle of the gauges, the construction of the tubular bridge, and the completion of the
high level bridge.

It is impossible to record the labours of the engineer
during the interval between the opening of 1844 and
the close of 1850 without contrasting them with the in-
trigues of adventurers who regarded railway enterprise
as gamesters regard a gambling table.   The triumph of
these adventurers was brief.   Just as the worker reached
the fullness of his fame, the chief speculator dropped
from his eminence, to be scouted by those who had
fawned on him in prosperity, and to be despoiled by
those whom he had benefited even more than by those
whom he had wronged.

The rest of this memoir will be devoted to a consideration of Robert Stephenson's great public parliamentary contests, in connection with the atmospheric system
and the gauges; to a description of those remarkable
achievements by which he will be known as the 'builder
of iron bridges,'—and to a general view of his professional

and personal history from the time of his entrance into the House of Commons as member for Whitby in 1847 up to the time of his death eleven years afterwards.

But before this second portion of Robert Stephenson's life is entered upon, in order that the reader may have a complete picture of the movement which he influenced, it will be necessary to glance at the history of railway enterprise and railway legislation.

CHAPTER XIII.

RAILWAY PROGRESS AND RAILWAY LEGISLATION.

First Act of Parliament authorising the Construction of a Railway—
Railway Developement from the Year 1801 to 1846 inclusive — The
Railway Mania of 1825-26 — The Railway Mania of 1836-37 —
The Railway Mania of 1845-46—Difference between the Crises of
1825-26 and 1836-37 and of 1845-46 — Report from Committees,
1837—Bubble Companies—Parliamentary Influence—Parliamentary
Corruption— Compensation; Stories of — The Parliamentary Com-
mittee as a Tribunal — Robert Stephenson's Views on Parliamentary
Legislation — Observations on his Project for a ' Preliminary Board
of Inquiry ' — Causes of Parliamentary Inconsistency — Stories of
the Parliamentary Bar — Professional Witnesses in the House of
Commons : Robert Stephenson, Brunel, Locke, Lardner, Bidder —
Great Britain compared with other Countries in respect of Railway
Developement—Results—Proposal for Railway Farmers—Proposal
for a Railway Bank.

RAILWAY organisation, like most important commercial systems, was an affair of small commencement; and to this fact can be traced the principal defects
and errors of railway legislation. The early tramways
were private works, undertaken at the sole cost, and
carried out for the benefit of private traders who for
generations bought ' way leave ' of landed proprietors,
and occasionally made arrangements of cooperation with
the owners of adjacent roads without seeking parliamentary sanction. It was not till the middle of the last
century that the legislature was first solicited to authorise the construction of a railroad, and so received a

first instalment of that business which, during the last forty years, has swelled to a prodigious bulk. A private act of the 31st Geo. II. (1758) has reference to the road used for coal carriage to Leeds, on which Blenkinsop's patent locomotives used to run, with toothed driving-wheels working on a rack-rail. Clauses are also found in many of the early canal acts, empowering the proprietors of the canals to construct railways in connection with their water ways. The first year of the present century, however, saw the railway instituted in this country as a means of public convenience. In 1801 the Surrey Iron Railway Company was incorporated, with power to construct an iron tramway for public use. A survey of the following table will show the course taken by railway enterprise, until it became one of the greatest and most complicated of existing commercial interests :—

	Acts empowering the Construction of New Lines	Acts amending Provisions and enlarging Powers determined by previous Acts	Total
1801	1	—	1
1802	2	—	2
1803	1	—	1
1804	1	—	1
1805	—	1	1
1806	—	2	2
1807	—	—	—
1808	1	—	1
1809	3	—	3
1810	1	1	2
1811	3	1	4
1812	2	1	3
1813	—	—	—
1814	1	1	2
1815	—	1	1
1816	1	—	1
1817	1	—	1
Carry forward	18	8	26

	Acts empowering the Construction of New Lines	Acts amending Provisions and enlarging Powers determined by previous Acts	Total
Brought forward	18	8	26
1818	1	—	1
1819	1	—	1
1820	—	1	1
1821	2	1	3
1822	—	1	1
1823	—	1	1
1824	2	1	3
1825	8	1	9
1826	10	1	11
1827	1	5	6
1828	5	5	10
1829	5	4	9
1830	5	3	8
1831	5	4	9
1832	5	4	9
1833	5	6	11
1834	5	9	14
1835	8	11	19
1836	29	6	35
1837	15	27	42
1838	2	17	19
1839	3	24	27
1840	—	24	24
	135	164	299

This table is a concise epitome of the history of railway enterprise during the forty years to which it refers. The first twenty-four years saw exactly the same number of acts passed. In 1825, however, a sudden start was made in consequence of the growing confidence in the Stockton and Darlington line. In the following year, when the success of that undertaking had been ascertained, the number of bills for new lines was ten. The commercial trouble of 1826 reduced the number of bills passed in the following session to one. In 1828, however, a fresh start was made, and steadily maintained till 1836, when the *first* great railway mania reached its height, and gave the public in the course of the session no less than twenty-nine new bills. In 1837 the first great mania began to subside, just as the works of the London and Birmingham line (to which the mania was in a great measure due) were on the eve of completion, and the passion for railway speculation was for a time so much suppressed, that the years 1838 and 1839 saw only five bills for new lines passed, and the year 1840 did not see even one. The lull, however, was only the precursor of a storm, the fury and ruin of which made the madness and misery of the railway mania of 1836 sink into insignificance. Robert Stephenson's London and Birmingham line had familiarised the London public with railways, and its success was a constant witness in support of those ambitious speculators who are always eager in exhorting the industrious and thrifty to find for their savings a better investment than the public securities. In 1841 an attempt was made to set the ball rolling once more, and a bill was granted for the construction of the Hertford and Ware branch; but so little was the country as yet in humour to renew the ruinous game of 1836, that even this little branch, five miles and three quarters in length, was not constructed. The session of 1842 saw the advent of George Hudson to London, and bills passed for the Newcastle and Darlington line (about which it will be necessary to speak more fully hereafter) and a few branch lines. The depression still continued. The parliamentary year of 1843 saw little that was new in the way of railway projection. But in the next session the floodgates were opened, and the deluge commenced which in three short years enriched rogues, beggared honest men, swept away the savings of sober industry, and reduced countless families to destitution. In 1844 bills were granted for the construction of forty-eight new lines, extending over 700 miles, at an estimated expenditure of £14,793,994. The allowance for 1845 was 120 new lines, measuring 2,883 miles, at a computed cost of £43,844,907. In the following year (1846) legislative liberality went so far as to authorise the expenditure of £121,500,000, on *two hundred and seventy-two new lines*, covering *four thousand seven hundred and ninety miles*. In all, the amount of the national wealth assigned in these three sessions of Parliament to railway enterprise was *one hundred and eighty millions, one hundred and thirty-eight thousand, nine hundred and one pounds*.

So long as applications for new lines were few, a parliamentary committee was the best possible tribunal for deciding on the propriety of investing private individuals with power to construct the required lines. From 1801 to 1824 inclusive, Parliament (as has been seen) granted only one bill per annum. Whilst the concession of one act a year for the construction of a small road for the convenience of local commerce was enough to satisfy the public demand for railways, there were no grounds for suspecting that assemblies, which had already considered the claims of canal-owners and projectors of public roads and bridges, would be found incompetent to decide with wisdom and equity on cases connected with the creation of public tramways. Now and then the projectors of an iron road between a nest of collieries and a neighbouring port might possibly be defeated in their application to Parliament, through the

interest of local members, but such interference would
not be likely to be either frequent or of permanent effect;
and even in the very few cases where local interests
might steadily and triumphantly combine against the
public good, the victims of such combination would be so
few, and so exceptional, that the nation at large could not
be expected to pay them much heed.

It is no purpose of the present work to collect materials
out of which the malignant might frame charges of corruption against individuals; but it is necessary to give a
truthful picture of evils which arose only a few years
since from circumstances peculiarly unfavourable to
disinterestedness and integrity. Until the public awoke
to a full sense of the benefits of the railway system, they
were slow to discern the injustice and evil consequences
of allowing members of the legislature to sit in judgement
on cases affecting their private fortunes. Indeed, far
from dreading, they found pleasure in calculating,
that the decisions of committees would be given in
accordance with the selfish instincts of the individuals
composing those committees. So universal amongst all
classes of society was the antagonism to railways, from an
apprehension that they were injurious to vested interests,
that gentle and simple viewed with equal complacency
the constitution of tribunals which necessarily sympathised
in a very high degree with the prevailing prejudice. At
first, therefore, as applications to Parliament for public
railways increased in number, the public felt that the
general interests of property were secured by the conclusions of railway committees composed of the persons
through whose estates the projectors wished to carry
lines. When it was ascertained that the opposition of

members was removed by pecuniary consideration, the
moral sense of the country, far from being shocked at the
corruption, gave it their sanction. The enormous sums
that railway companies had to pay in complying with the
required forms of parliamentary application, and the yet
more exorbitant sums that had to be expended in buying
off (under title of ' compensation ') the opposition of influential proprietors, appeared to the general public in
the light of guarantees that old interests would meet with
extreme consideration from the new innovators. The
publicity with which demands and proposals and arrangements, having compensation for their object, were made
to railway companies, by itself shows how these bargains
were regarded by the community at large. An impoverished nobleman, owning a house and park (of the
value of £30,000) in a county through which one of the
earliest railways was carried, for a small strip of his park,
occupied by the railway, which ran quite beyond the
sight-range of his windows, obtained no less a sum than
£30,000—or the entire value of the estate which the line
was supposed only to *depreciate*. A few years afterwards
this same peer sold another corner of the same park for
another line for a second £30,000, and when he had thus
extracted from two Companies £60,000 as compensation
for damage done to his estate, the original property was
greatly augmented in value by the lines which, it was
represented, would inflict upon it serious injury. Of
course it was well understood that two sums of £30,000
did not represent the price of the land, but the price of
the peer's parliamentary interest.

Similar cases were of constant occurrence ; and, far
from rousing public indignation, they met with public

approval. *Any* amount that could by *any* means be squeezed from the funds of a railway company under the name of compensation public opinion decided to be legally and honourably acquired. As compensation for 'severance'—i.e. for the injury presumed to be done to an estate previously lying within a ring fence—a proprietor (after requiring that bridges should be built at so many points of the line that 'the severance' would practically cease to exist) would demand two, three, or four thousand pounds, in addition to the extortionate price already paid for the land actually given up to the line. It was to no purpose that the agents of railway companies demonstrated that this 'severance' was merely an imaginary grievance, and effected no real injury to the estate. Refusing to see the question in this light, the owner remained steady to his demand, and gained his 'severance' compensation. Having thus sold a strip of land at four, five, or six times its value, and obtained heavy compensation for the purely imaginary grievance, the owner would then candidly avow that 'the severance' of his land caused him so little discomfort that he could do with only half or a quarter of the stipulated bridges, and that he would for a further sum free the company from the obligation to build the unnecessary bridges. In the early days of public railways, companies were powerless to resist such extortions. They had to buy in hard cash the goodwill of the community. Frequently the owner who drove the hard bargain was a peer, or a member of the House of Commons, and had interest enough at Westminster to effect a combination that would upset the bill for the proposed line before committee. In other cases he was allied by blood or

friendship to county magnates who had such influence; or
even where he was only a wealthy yeoman farmer, he
often had sufficient local power to rouse the opposition
of surrounding owners, who felt they had a common
interest to serve in plundering the new railway companies.

The result was that in too many cases a bill was
obtained for a new line on grounds altogether distinct
from its merits; and in an equal number of cases a
line (like that of the London and Birmingham) based
on the soundest commercial policy, and demanded by
national interests, failed to win parliamentary sanction,
because it disturbed the operations and broke into the
property of a few private persons.   In due course,
however, a change was wrought in public opinion.   The
utility of railways, and the benefits conferred by them
upon the entire community, having been demonstrated by
experience, the impropriety was seen of permitting
railway questions to be decided by persons who were
immediately and personally interested in them.   It was
perceived that a needy member of Parliament, who had
been offered £5,000 for a strip of land not worth £500 by
the directors of a projected railway, was as little likely
to be solely guided to his decision by the actual merits
of the proposed line, as any impoverished member of the
judicial bench would be likely to hold his ermine spotless
if a similar bribe were offered, under circumstances that
secured him from exposure.   A parliamentary resolution,
therefore, excluded from the committee sitting on any
proposed line all members who either held land through
which the line was to run, or were otherwise commercially
interested in the ejection or passing of the bill.   This

measure of reform did much to check the scandalous traffic of parliamentary influence, which had been previously carried on, without even a pretence of conceal-ment or shame, by members of both houses of legislature. But it by no means put an end to all corrupt practices; and indeed the principal evil at which it was directed had in a very great measure ceased to exist, and given place to another form of legislative abuse.

That the reader may understand this, it is necessary that he should survey the course of railway enterprise from another point of view.

The table and *resumé* given at the commencement of this chapter show that between 1801 and 1846, inclusive, there were three separate periods when speculation in railways made a great start,—each start followed by a corresponding collapse. The first of these periods was in 1825 and 1826 (in a great degree induced by the operations on the Stockton and Darlington line); the second was in 1836 and 1837 (when the first metropolitan locomotive railway was near completion); the third was in 1845 and 1846, when George Hudson, at the height of his success, had for more than three years been lead-ing the country to believe that 'management' was the only thing required to make any line of railway answer. Business men who can recall from personal experience the events of these three crises sometimes designate them the *three periods of railway mania.* The first crisis, how-ever, was so slight as compared with the second, and the second was so slight as compared with the third, that some persons speak only of two important paroxysms of railway gambling, whilst with a great majority of English-men the almost universal madness of 1845 and 1846 is *the* railway mania, and *the only* railway mania worthy of record.

Between the crises of 1825 and 1836, and the mania of 1845, there was as wide a difference in character as in magnitude. At the two former periods the speculators were for the most part obscure adventurers : whilst, in the last outbreak, the gamblers comprised every rank of society, and embraced a greater proportion of the aristo-cratic and educated classes than of the lower. In 1825 and 1836, railways were still regarded by nine-tenths of the inhabitants of Great Britain as inventions that could never benefit society. Landed proprietors, from the peer to the petty yeoman, and all their dependents, viewed them with either distrust or violent hostility. The great monetary chiefs of the kingdom also opposed them. In London, with the exception of a few such men as Mr. Richardson and Mr. Glynn, there was scarcely a banker or eminent broker who did not rank railway speculation with the South Sea bubble. In Durham Mr. Pease, familiar from boyhood with the railroads of the northern coal-field, advocated the cause of iron roads ; but in Norfolk, where such roads were unknown, a wealthy banker was foremost amongst the opponents of railways. In Liverpool the new road was appreciated ; but in the South, at all the principal seats of learning and commerce, it was decried on every consideration of policy. A banker (whose name it would be unfair to mention in connection with this story), residing in one of the Eastern Counties, even went so far as to make a will, leaving in the hands of trustees a considerable property to be expended on parliamentary opposition to railways. It should be added, that the worthy gentleman who made this preposterously absurd disposition of his estate lived to see his folly, and devote his wealth to better purposes.

In 1825 and 1836, the multitudes ready to embark in railway speculation comprised comparatively few monied persons. In the latter period, there was a crowd of projects, and there was a mob of shareholders; but in a great majority of cases, the schemes and the projectors wanted alike the countenance of tried engineers, and the support of solvent speculators. It was a time very different from the crisis ten years later. Applications were made to parliament for new lines, of which the engineers were charlatans, incapable of taking the level of a grass-plot, the directors were unknown clerks, and the shareholders were little more than beggars. In some instances plans were submitted to Parliament, the engineers and draughtsmen of which knew the country concerned in them only through ordnance maps. Mr. Cundy's London and Brighton line was one of several similar efforts. If they were not amply attested by evidence recorded in parliamentary blue-books, a reader of the present day could scarcely credit the stories to be told of the mushroom companies of 1836. An attorney without practice, a few bankrupt traders, and as many brokers expelled from the Stock-Exchange, would hatch a scheme for a new line. The attorney (invariably at the bottom of the mischief) undertook the legal business of the association; another of the party, without any regard to his previous education, started as the engineer; a third secured for himself the post of secretary; whilst the rest of the conspirators consented to be nothing more than directors, with handsome fees to be paid out of the first money acquired as 'deposit' on shares taken by their victims. A prospectus was speedily concocted and a sham survey made. The principal business of the first few months was to find shareholders. To draw dupes it was necessary to have a show of business, and to display a handsome list of subscriptions. This the agents of the company effected by getting signatures from discharged bank-clerks, insolvent schoolmasters, touters of the Stock-Exchange, assistants of sheriff's officers, hotel-waiters,* cab-drivers, keepers of houses of ill fame, and persons unable to keep a house of any kind whatever. Men whose names were entered on lists as shareholders to the amount of thousands, and who were represented as having paid 'deposit' money to the amount of many hundreds of pounds, acknowledged on examination before parliamentary committees, that at the dates of their respective signatures they had not a sixpence in the world, did not know where to look for a dinner—had not a vocation whereby they could earn an honest subsistence. The mode by which these indigent knaves were induced to sign the subscription lists was not less remarkable than their fraudulent impudence. A gentleman would meet them as they hung about the purlieus of Capel Court, waiting to run errands or discharge commissions for chance employers, and would inform them that there was *a petition* being signed in a certain house at a certain street, and that every signer of the petition would receive ten

---

* Vide 'Reports from Committees, 1837: First Report from Select Committee to inquire into the Matters of several Petitions complaining of the Names of certain needy and indigent Persons having been inserted in the Subscription Lists of several Railways [Deptford and Dover], with Minutes of Evidence and Appendix. Second Report, same [Westminster Bridge, Deptford, and Greenwich], with Minutes of Evidence and Appendix. Third Report, same [City, or Southwark Bridge and Hammersmith], with Minutes and Appendix.'

shillings and sixpence — *not for his signature,* but *for his trouble in going to the appointed house for the purpose of signing.* Induced by such representations these fellows went to the office, wrote their names down on the subscription lists, subscribing for shares as if they were millionaires. As they quitted the office they each received from an unknown agent in a dark passage the price of their labour — that is to say, *their trouble* in coming. In examination before committees these men did their best to secure the ' agents from detection. The person who paid them was of course quite unknown to them; and the passage in which they were paid was of course so dark that it was utterly impossible for them to distinguish the features of their benefactor, and, equally as a matter of course, they were under the impression that, in signing the subscription, they were acting usefully and honestly. Sometimes these ten-and-sixpenny capitalists were at a loss how to describe themselves, and forgot to put ' Gentleman ' or ' Esquire ' after their names. The secretary, however, easily rectified that slight omission. On other occasions they exhibited hesitation or ingenuity in assigning to themselves reputable residences. One subscriber wrote himself down a resident of a well-known and respectable street, because he had formerly lodged in it; and another capitalist described himself as a householder in a good square, because he often took a walk in the neighbourhood. Such men were good enough to serve the purpose of the unscrupulous agents who paid for their services. The subscription lists were seen to be full of names: the numbers of shares subscribed for, and the amounts of money deposited, were quoted in the organs of railway intelligence. Shares mounted to a premium, and credulous dupes rapped at the doors of the bubble companies, anxious to become *bonâ fide* purchasers of stock. In the ensuing parliamentary session, the bills of these fictitious associations were thrown out on examination of their merits, or summarily dismissed for non-compliance with standing orders. But in the meantime the deposit money, and sums paid for shares transferred at premium, had passed from 'the sheep' into the hands of their fleecers. In more than one case, a company, together with its office, directory, and agents, vanished before the commencement of the parliamentary session; and when its victims made anxious enquiries after their defrauders, they learnt that the directors were the scamps of city cliques, and that 'the office' was nothing more than a room hired by the week.

Whilst railway projectors numbered such scoundrels amongst their ranks, and whilst such practices were of frequent occurrence in the transactions of the railway market, the public had some excuse for looking complacently on the selfish policy of members of Parliament. It was argued, not without reason, that the heavy exactions to which *bonâ fide* railway companies had to submit before they could carry out their purposes, were at least some guarantee that their promoters were not mere penniless knaves bent on robbing the public. It would have been well, however, if parliamentary corruption had been confined to such extortion as was covered by the word 'compensation.' Unfortunately for the national character, there were members of the legislature who systematically sold their parliamentary interest for money considerations, in the manner of those representatives of the United States who are known to be accessible to 'lobby influence.' The

time has not yet arrived when it would be right to speak
fully on this point. Possibly some future Pepys' diary
will reveal to Englishmen of the twentieth or twenty-first
century the names of those British senators of the past
generation who gave their votes for gold, and will describe
minutely the exact circumstances of particular compacts.
For the present, it is enough to state the *fact*—which is
too important, as an indication of social morality, to be
altogether passed over without mention. Nor need any
member of the existing legislature deem the honour of
his order attacked by these remarks, for as far as the
materials used for this work throw light upon a dis-
agreeable topic, it can be stated that without exception
the men who profited by such shameless corruption have
disappeared from public life.

Before quitting this painful part of an important subject,
it ought again to be impressed on the reader that railway
companies were subjected to extortion alike by all ranks
of society. When a railway passed through a provincial
town, its directors found the demands of merchants and
petty traders quite as exorbitant as those of the landed
aristocracy. Mr. Bidder's experience as engineer of
the Blackwall line, under George Stephenson, gives
emphatic support to this statement. The Blackwall line
was the first railway to pass through a very populous
suburb and a crowded quarter of the metropolis; and
in completing that important work the directors had
daily to submit to demands for compensation, compared
with which the exactions of county gentry were liberal
arrangements.

By 1845, it was found that railways did not depreciate
the property, lower the rents, scare the cattle, or poison

the atmosphere of the districts they traversed. It was even seen that, morally and physically, the condition of the humble classes was improved by the means, with which railways presented them, of quitting over-crowded neighbourhoods, and seeking employment where labour was in demand. Instead of dying from frenzy, or catching disease from the waste steam of the locomotives, the live stock of distant counties also derived benefit from the change. Fodder of superior quality and diminished price was conveyed to them by the goods-trains, and breeds were improved by the greater facility with which agriculturists could procure stocks from remote counties; and in addition to the benefits thus conferred on commerce and the working classes, the convenience of the new method of transit was highly appreciated by the wealthy. Gentlemen who, like Mr. Assheton Smith, wished to represent their shires in the House of Commons, and at the same time hunt their fox-hounds two or three days a week, soon learnt to approve a system which brought the best hunting countries within two or three hours' ride of the capital.

In 1845 the aim of the aristocracy, therefore, was to obtain the greatest possible number of iron-roads, and to have them running close to their front doors. There was no reason to fear that they would not pass good bills. The evil was that, in their anxiety for railroads, they passed bad ones also. Formerly railway projectors—by 'compensation,' and other forms of bribery—used to purchase the good-will of a party within the legislature. In 1845 corrupt action went on, but in a different manner. Railway projectors (the corrupting power) were no longer outside the walls of the Houses, but within them. Peers

and members of the Lower House were avowedly engaged
as traffickers in the railway market, their names being
advertised in every quarter as promoters or directors of
lines. One consequence of this was the comparative
impotency of the rule which forbade members to sit in
committees on lines in which they were personally
interested. Members attached to the 'railway interest'
voted for each other's projects. A sat in committee and
voted for the line in which his friend B was personally
interested; and B in like manner watched with paternal
care over the parliamentary career of the line in
which A was personally interested. The results of this
system of amicable cooperation were (as has been already
seen) a hundred and twenty new bills in 1845, and two
hundred and seventy-two in 1846. Indeed there was in
those years scarcely a single person, in either the House of
Lords or the House of Commons, who was not, personally
or through his connections, anxious that a bill should be
obtained for some particular new line.

In the crisis of 1836, and also in the crisis of 1845, the
parliamentary committee was a tribunal ill-constituted to
do justice between railway projectors and the public; but
it must be acknowledged that under the circumstances it
would have been extremely difficult, if not impossible, to
devise a better court of enquiry. Robert Stephenson was
always a strong advocate for the creation of a railway
board, composed of persons specially qualified by education
to preside over railway legislation and administration. But
it is open to something more than doubt whether any court
that could have been formed to carry out his views would
in its practical working have been more efficient, or pure,
than the parliamentary system, with all its shortcomings,

blunders, and inconsistencies.   In an address delivered
to the Civil Engineers, on taking possession of the
Presidential chair at the Institution in 1856, Robert
Stephenson observed :

Little more than a quarter of a century has elapsed since
Parliament first began to legislate for railways.   In that period
a multitude of laws have been placed upon the statute-book
which will certainly excite the wonder, if they fail to be the admiration of future generations.   The London and North-Western
Railway alone is regulated, as is shown by a return of Mr.
Hadfield's, by no less than 168 different Acts!   Of these the
greater part were passed in the present reign.

But it is not so much the number of the statutes regarding
railways that excites surprise.   The extraordinary features of
the parliamentary legislation and practice consists in the
anomalies, incongruities, irreconcilabilities, and absurdities which
pervade the entire mass of legislation. . . . . . . Not
only is the legislation irreconcilable, but throughout the
quarter of a century during which attention has been given to
this branch of legislation, the Acts of Parliament have been
wholly at variance with its own principles.   To illustrate this:
several different select committees have, at various times, deliberately reported against the possibility of maintaining competi-
tions between railways, and to this principle Parliament has as
often assented.   Yet the practical operation of the laws which
have received legislative sanction has been throughout, and at
the same time, directly to negative this principle, by almost invariably allowing competition to be obtained, wherever it had
been sought.   Parliament has therefore been adding to the
capital of railway companies, whilst it has been sanctioning
measures to subdivide the traffic.   The decline of dividends was
an inevitable consequence.

Again, in 1836, the House of Commons required its committee
upon railway bills specially to report as to the probability of railways paying.   This principle has, however, been gradually departed from, until such enquiry is now considered and treated as
unimportant.   Legislative sanction having been given to a line,
it might be supposed that Parliament would also grant adequate

protection, exacting from the railway public facilities and advantages in return for the rights afforded to it. Whilst the legislature and the government have exacted facilities and advantages even beyond what they had a fair right to demand, so far from protecting the interests of those to whom they conceded the right, they have allowed—nay, they have encouraged—every description of competition. What has been the result? As regards the completeness and perfectness of the line first made, obviously it must have been most injurious; as regards the interests of the shareholders, no doubt it has been, in many cases, most disastrous. But how does the case stand as regards the public? Why, whatever may have been the effect for a time, the competition which Parliament has permitted has invariably been terminated by combination, so that the public have been left precisely where they were.

But the incongruities are by no means the worst features of the parliamentary legislation now under consideration. Mr. Hadfield's return has been spoken of. That return — in itself exceedingly incomplete, and affording no information of any sort respecting forty-five railway companies, for which Acts have been obtained — shows that the amount expended by existing railway companies in obtaining the Acts of Parliament by which they are empowered has been no less, in parliamentary, legal, and engineering costs, than fourteen millions sterling. No sooner was that fact placed on record, than a universal outcry burst from the alarmists. 'See,' it was said, 'how shareholders have been plundered; see how their money has been squandered; look at this vast amount of waste, and consider how much better it would have been in your own pockets!' But in no one case did those who made these bitter comments attribute the monstrous result to the proper cause. Railway directors and officials have been held responsible for what has been the fault, solely and exclusively, of Parliament itself. What interest can directors and officers have in group committees, wherein counsel must be fee'd for attendance during, perhaps, ten or twenty days when they are never heard nor wanted. What interest can directors or officers have in keeping crowds of witnesses in London, at great expense, awaiting the pleasure of a committee, which is engaged upon another measure, and which can rarely foresee or indicate when those witnesses will be required. The

ingenuity of man could scarcely devise a system more easy than
that of getting a railway bill through the legislature. But who
devised that system?—Parliament itself. Who have begged, and
prayed, and implored for alteration unavailingly?— directors and
officers of companies. An illustration may show more graphically
how Parliament has entailed expense upon railway companies,
by the system it has set up. Here is a striking one. The Trent
Valley Railway was, under other titles, originally proposed in the
year 1836. It was, however, thrown out by the Standing Orders
Committee, in consequence of a barn, of the value of about 10*l.*,
which was shown upon the general plan, not having been exhibited upon an enlarged sheet. In 1840 the line went again
before Parliament. It was proposed by the Grand Junction
Railway Company (now part of the North-Western). No less
than 450 allegations were made against it before the Standing
Orders Committee. The sub-committee was engaged twenty-
two days in considering those objections. They ultimately
reported that four or five of the allegations were proved; but
the Standing Orders Committee, nevertheless, allowed the bill
to be proceeded with. Upon the second reading it was supported
by Sir Robert Peel, and had a large majority in its favour. It
then went into committee. The committee took sixty-three days
to consider it, and ultimately Parliament was prorogued before
the report could be read. Such were the delays and consequent
expenses which the forms of the House occasioned in this case,
that it may be doubted if the ultimate cost of constructing the
whole line was very much more than the amount expended in
obtaining permission from Parliament to make it.

This example will show the delays and difficulties with which
Parliament surrounds railway legislation. Another instance will
illustrate the tendency of its proceedings to encourage competition. In 1845 a bill for a line now existing went before
Parliament with no less than eighteen competitors, each party
relying on the wisdom of Parliament to allow their bill at least
to pass a second reading! Judged by such a case, the policy of
Parliament would really seem to be to put the public to expense,
and to make costs for lawyers, and fees for officers. Is it possible
to conceive anything more monstrous than to condemn nineteen
different parties to one scene of contentious litigation? Bear
in mind that every additional bill received by Parliament entailed

additional expense, not only on the promoters of that one bill, but on all the other eighteen competitors. They each and all had to bear the costs, not of parliamentary proceedings upon one bill, but of the parliamentary proceedings on nineteen bills. They had to pay, not only the costs of promoting their own line, but also the costs of opposing eighteen other lines. And yet, conscious as Government must have been of this fact, Parliament deliberately abandoned the only step it ever took, on any occasion, of subjecting railway projects to investigation by a preliminary tribunal.

After glancing at the facilities afforded by Parliament to landowners for demanding exorbitant compensation, so that ' of the £286,000,000 of railway capital expended, it is believed that nearly one-fourth has been paid solely for land and conveyancing,' Robert Stephenson went on to suggest, as a remedy, ' a tribunal competent to judge and willing to devote its attention to railway subjects only.'

' What we ask,' he said, ' is knowledge. Give us, we say, a tribunal competent to form a sound opinion. Commit to that tribunal, with any restrictions you think necessary, the whole of the great questions appertaining to our system. Let it protect private interests apart from railways; let it judge of the desira- bility of all initiatory measures, of all proposals for purchases, amalgamations, and other railway arrangements; delegate to it the power of enforcing such regulations and restrictions as may be thought needful to secure the rights of private persons, or of the public; devolve on it the duty of consolidating, if possible, the railway laws, and of making such amendments therein as the public interests and the property now depending upon the system may require; give it full delegatory authority over us in any way you please. All we ask is, that it shall be a tribunal that is impartial, and that is thoroughly informed ; and if impartiality and intelligence are secured, we do not fear the result.'

It is here seen that the chief charges preferred against parliamentary legislation by Robert Stephenson are those

of inconsistency and inordinate cost. Without a doubt
the accusations were fully sustained by facts; but the
faults complained of would unquestionably have disfigured
the operations of any other system.

The evils of parliamentary legislation were evils
necessarily consequent on a free system of commercial
enterprise and social developement. A paternal government might have mapped out the United Kingdom (as
paternal governments did *subsequently*! map out some of
the continental countries), and have declared what towns
and districts should enjoy railway communication—and
under what conditions. Such legislative interference
would unquestionably have given us railways at a very
much lower cost ; and millions of money would have remained undisturbed which under existing circumstances
changed owners ;—but as certainly we should in the long
run have been less liberally supplied with roads.

Much of the exorbitant expense of our railway contests
must be set down to the fact that Parliament was suddenly
inundated with railway business, to discharge which it
was unprovided with fit machinery. Business which had
previously been an occasional feature of parliamentary
enquiry became the chief subject of attention throughout
the session. The consequent confusion was not confined
to Westminster. Throughout the country there was such
a demand for surveyors and draughtsmen, that mere
mechanics could earn incomes that seldom reward the
ordinary members of the learned professions. The few
barristers who had the confidence of the few parliamentary agents who managed the private business of the
House of Commons leapt into the enjoyment of fees
which, at this date, seem fabulous. Their work was

comparatively simple, and required only a slight knowledge of law. At first, they fought on the merits of lines.
In the Liverpool and Manchester, the London and Birmingham, the Blackwall, the London and Brighton, and
similar contests, counsel, under the guidance of engineers,
worked up the estimates, mechanical difficulties, and
scientific problems of the undertakings. But with the
rapid increase of business they paid less attention to the
merits of projects, and procured the triumph or ejection
of bills by attention to 'the forms' required by committees.

Pages could be filled with anecdotes of great fees paid
to favoured advocates for very little work. In the full
height of the mania, one of them was found under the
shady trees in St. James's Park, leisurely lounging about
and feeding the water-fowl with crumbs of biscuits, before
the afternoon had scarcely begun. 'Why!' cried the
friend, 'how come you to be here at this time of day?'—
'Oh, my dear fellow,' was the naïve answer, 'I am
engaged to appear before nine different committees at
this very hour; and as it is impossible for me to accomplish that, I thought I would come out here, and enjoy
myself. Each of my clients, when he finds I don't appear
in his committee-room, will suppose I am talking away in
another. So it will be all right.'

Not less overworked, though by no means so highly
paid, were the engineers employed to give evidence before
committees. Of those who thus distinguished themselves,
several have already gone to another world. Robert
Stephenson, Brunel, Locke, and Jacob Samuda are no
more; Mr. Bidder and Mr. Vignoles remain. At first,
the evidence of the engineer was always given in

strict good faith ; and such, to the last, was the case with
Robert Stephenson and engineers of the same moral
stamp. Other gentlemen, however, with more elastic
consciences, regarded their position as identical with that
of the advocate, and gave testimony on one side or the
other, just as they were paid. The ordinary fee to a
scientific witness before committee being ten guineas *per
diem* (whilst waiting to give, as well as whilst giving his
evidence), an engineer in great request as a parliamentary
witness could frequently during session pick up 100
guineas a-day in the various committee-rooms.

Robert Stephenson resolutely refused to give purely
venal testimony, or as a witness to say one word which he
did not conscientiously believe. At the opening of the
session of 1845, when the last great mania was advancing to its most violent phrenzy, railway companies even
went so far as to send to eminent engineers retaining fees
for their services in committee-rooms. One morning
Robert Stephenson and Mr. Bidder received, through the
post, cheques from various companies, amounting to more
than £1,000. No previous overtures had been made by
the directorates who forwarded these fees. Similar sums
were sent round, in like manner, to other leading engineers.
Both Robert Stephenson and his friend lost no time in
returning the cheques, with intimations that their evidence
was not a power to be bought and sold. To the honour
of their profession it may be stated that other engineers
acted in the same way.

Prominent amongst scientific witnesses during the
memorable railway contests was Dr. Dionysius Lardner,
who detected the importance of railway enterprise at
its first outset, and zealously interested himself in its

developement. His first separate work on the subject was
a collection of some 'simple rules,' by which parliamentary committees as well as capitalists might be guided in
forming an estimate of a proposed road. Formed on the
dawn of railway practice, the opinions of this brochure
were unsound. The Doctor had himself the good sense to
repudiate his raw theories. But unfortunately the 'litera
scripta' of his fallacious rules outlived their influence,
and day after day was his luckless work quoted against
him, to the infinite amusement of the parliamentary bar,
and amidst the suppressed titters of *practical* engineers who
were only too well pleased to behold the discomfiture of
the man of theory. Mr. Bidder was the mathematical
witness usually opposed to the Doctor. As soon as
the Doctor was called for examination, Mr. Bidder would
rise, and present the counsel, whose duty it was to cross-
examine the popular mathematician, with a handsomely
bound copy of the 'rules.' Instantly a titter would rise
amongst the *habitués* of the committee-rooms. Dr. Lardner
would turn crimson with irritation, and strangers would
be at a loss to surmise what was going on beneath the
surface. In another minute a barrister, smiling ami-
ably, would begin by asking—'I believe, Dr. Lardner,
you are the author of a little work containing a few "simple
rules" on railways?' 'Yes, sir; you know all about
that, sir,' the Doctor would answer; 'you asked me all
about that ten times yesterday.' 'Ah, but, Dr. Lardner,'
the tormentor would continue, 'that was not before *this*
committee. Indeed, I must beg you to give me a little
information about your "rules."'

In judging Parliament for its shortcomings and errors
in respect of railways, sight must never be lost of the

difficulties under which it acted. And before the censor condemns the system which gave us our iron roads, he would do well to reflect that there is no country which is so well supplied with the new means of locomotion — no country where iron roads are so plentiful, trains so numerous and rapid, and fares so low. The gravest fault committed by our system in its early career was extravagance, which reduced countless humble families to enrich a few great persons, or rear colossal fortunes for a few hundreds of attorneys and adventurers. In 1849 * the United States had, completed and in use, 6,565 miles of railway. Of this 2,842 miles (the expense of which is accurately known) cost £23,104,909, or about £8,129 per mile. It is assumed that the remaining 3,723 miles were made at the same rate of cost: the expense of 6,565 miles may be computed £53,386,885 ; whereas the 5,000 miles of railway in the United Kingdom cost £200,000,000. The difference of these two last sums is so vast that, after making every allowance for the cheapness of land in America, and the presence in the United States of numerous inducements to, as well as facilities for, the construction of railways, no impartial observer can be otherwise than struck with the scandalous prodigality of British expenditure. Still it must be remembered that we were the pioneers in railway developement, being the first to test the merits of the system. Profiting by our experience, other countries avoided our mistakes.

The American States have a greater length of railway

---

* Railway Economy: a Treatise on the New Art of Transport, its Management, Prospects, and Relations, Commercial, Financial, and Social. By Dionysius Lardner, D.C.L., &c. Taylor, Walton, and Maberly.

than ourselves, both actually and proportionately to their population. But from her extent of territory, America is of course by no means so well supplied with lines as Great Britain. And by every other mode of calculation —by proportion of length of railway to extent of territory, by proportion of railway capital to the population, or by proportion of railway capital to extent of territory— Great Britain is richer in railways than any nation of the world.

It may not, however, be presumed that our railway system is incapable of improvement. An instructive writer (Mr. W. Bridges Adams) has recently exposed its mechanical defects. In respect of management it has also grave deficiencies; and it seems scarcely credible that the present generation will pass away without making some attempts for their amendment. The best authorities on 'railway interests' (and the term includes the interests of the public as well as of shareholders) are unanimous in avowing the inefficiency of railway management by directorates elected from the shareholders. In the ruinous contests of rival lines, lowering their fares in the hope of reducing each other to bankruptcy, the incompetence of such controlling boards has been signally and frequently displayed. The competition of two railways working through the same tracts of country has on many occasions given rise to internecine war between their directorates. Increasing in vehemence, the commercial strife has degenerated into personal quarrel, and directors have not hesitated to sacrifice their dividends and embarrass their resources for the mere pleasure of inflicting injury on their antagonists. Such battles, regarded from one point of view, are amusing: but the gloomy reflection they suggest to prudent observers is, that the money squandered by directors is taken from the pockets of shareholders, who regard with dismay the policy to which they are sacrificed. A renewal of these exhibitions of folly would be obviated if companies, instead of working their lines themselves, would let them to farmers, who, like the farmers of turnpike roads and bridges, should pay a certain fixed or variable rent to the shareholders, and retain the surplus receipts.* If such a plan were adopted, a new class of business men would speedily arise, who would see their advantage in providing in the best possible way for the public convenience, and would be chary of engaging in contests, the entire cost of which would fall on themselves. By such a system, shareholders would be secure of their dividends, and the public secure of good accommodation. The only individuals who would suffer by the reform are the gentlemen who at present play with money which is not their own.

Besides this wise measure of amendment, a proposal for a railway bank, undertaking to discharge the functions of bank and also of clearing house to all the Railway companies of the United Kingdom, has of late been a frequent subject of discussion with the most influential personages of the railway market.

---

* This system has already been adopted with advantage on some few lines in Switzerland.

CHAPTER XIV.

THE ATMOSPHERIC SYSTEM OF RAILWAY PROPULSION.

Remarkable Episode in the History of Railways — Correction of
Nomenclature — Objects of this Chapter — General Modes of Loco-
motion—Constant rivalry between Locomotive and Stationary Steam-
power — Liverpool and Manchester Railway —Walker and Rastrick's
Report — Stephenson and Locke's Reply— Triumph of the Locomo-
tive — Renewal of the Stationary Plan in the Atmospheric form —
Early Inventors — Papin — Medhurst — Features of his Schemes —
Vallance — Pinkus — Clegg — Jacob and Joseph Samuda — Private
Experiments — Trial of their Plan on the Thames Junction Railway
— Description of the Apparatus — Proposal to apply it in Ireland —
Smith and Barlow's Report — Application on the Kingstown and
Dalkey Line — Arguments in favour of the Plan — Robert Stephen-
son's attention called to it in reference to the Chester and Holyhead
Railway — His Report — Public Interest excited — Croydon Railway
Parliamentary Committee — The Railway Mania — Appointment of
a Committee of the House of Commons to enquire into the Merits of
the Plan — Their Report in its favour — Culminating point of the
History — Contests in Parliament — Application of the Atmospheric
System in practice — Thames Junction Line—Kingstown and Dalkey
Line — Croydon Line — South Devon Line — Paris and St. Ger-
main Line — Summary of Results — Mechanical Efficiency —
Economy — General Applicability to Railway Traffic — Reasons for
its Abandonment — Conclusion.

THE attempt that was made some years ago to introduce the atmospheric system of propulsion upon railways, forms such a remarkable episode in their history,
that it deserves a somewhat extended notice.

The invention here referred to is often termed the
' Atmospheric Railway,' but this is a misnomer.   In the

economy of railways, whether scientifically or practically considered, it is always desirable to distinguish between what relates to the road itself, and what has more especially to do with the movement of the vehicles upon it. The design and construction of the road are matters quite distinct from the system of *haulage* used ; and the atmospheric plan, being simply a peculiar mode of producing locomotion, forms no essential part of the *railway* properly so called.

It is proposed to give in this chapter a short historical and descriptive notice of the Atmospheric System of Railway Propulsion, from its invention, through its season of popularity, to its ultimate abandonment ; and we shall dwell particularly on the part taken in its history by Robert Stephenson. He was from first to last its determined and consistent opponent; at one time almost standing alone, in the face not only of the promising appearance of the invention, but of the conscientious and powerful advocacy which it received from many of his most eminent brethren in the profession.

There are three modes by which locomotion is usually effected upon railways ; namely, by animal draught, by locomotive steam engines, and by stationary steam power.

The first was given up at a very early period, as inadequate to the requirements of railway traffic, and is now only used in exceptional cases ; but between the second and third, i. e. the locomotive engine and the various modes of applying stationary steam power, to which latter class the atmospheric system belongs, there has been an almost constant rivalry.

At the time when the Liverpool and Manchester
Railway was formed, both systems had been tried.
George Stephenson had introduced his locomotives on
the Stockton and Darlington and other lines; while the
plan of drawing the carriages along by stationary
engines and ropes had also been used extensively by
him, and was at work in many parts of the country.
When the railway approached completion in 1828, the
directors, having determined that some more efficient
power than horses must be used, consulted two eminent
engineers, Mr. Walker and Mr. Rastrick, as to which
application of steam power, the locomotive or the
stationary, would be preferable for the purposes of the
line.

These gentlemen, after studying the examples of the
two systems then in existence, gave their opinion that, if
it were resolved to make the railway complete at once, so
as to accommodate the full traffic expected, the stationary
system was best;—but that if any circumstances should
induce the directors to proceed by degrees, and to proportion the power of conveyance to the demand, then
locomotive engines would be preferable. In any case,
however, they considered it necessary that stationary
engines should be used on the two inclines at Rainhill and
Sutton, with gradients of one in ninety-six, to which they
considered the locomotive system inapplicable.

In reviewing the detailed facts and reasonings given
in the reports, it would appear that the principal, if
not indeed the only ground for the preference of the
stationary system was a saving of something over twenty
per cent. in the cost of working, maintenance, and interest of capital, which the referees considered would

accrue by its use. Other points of comparison are mentioned, but no very decided opinion appears to be expressed in favour of either plan, except for the reason above stated.

The referees, therefore, evidently laid themselves open to attack on their main ground—namely, their estimates —from any one who had sufficient knowledge of the working of the two systems to detect any errors into which they might have fallen; and the cudgels were soon taken up by George Stephenson in defence of the locomotive. He urged strongly its superiority, in numerous reports and at numerous meetings of the directors, until at length this body, influenced by his persistent earnestness and by the undeniable weight of his arguments, resolved to adopt it, and instituted a public competition to determine the best form of the machine, the result of which is well known.

An excellent resumé of the arguments used, on these occasions, by the great founder of the locomotive system is given in a tract published in 1830, under the joint authorship of ' Robert Stephenson and Joseph Locke, civil engineers.' Robert lent his father very active aid in this matter; and it is clear that he had, at this early stage of the controversy, thoroughly mastered the principles of the dispute, and acquired the strong convictions in favour of locomotive power which he afterwards so unflinchingly adhered to.

The joint essay attacks, with much force, the correctness of the estimates of the referees, asserting that, instead of the stationary system being the cheaper in the working, there would be an economy in favour of the locomotive in the proportion of eight to five. It points

out that the capabilities and advantages of the locomotive
system had not been properly appreciated, and that some
disadvantages of the rival plan had been overlooked ;
and it concludes with some able and far-sighted remarks
on the two systems generally, which, as bearing strongly
on the subsequent revival of the stationary plan in the
atmospheric form, it may be useful to reproduce here.

In drawing a comparison (say the authors) between locomotive
and stationary engines, the relative expense is certainly of vast
importance; but, though this is a primary object, that of despatch and public accommodation is of the utmost consequence,
and may be said to rank higher in the scale of importance than
expense, when the difference between the two systems in the
latter item is not very great.   When the traffic upon a railway
is either small or variable, the locomotive engines are not only
cheaper, but much more convenient, because the number of
engines in operation at one time may be regulated as the trade
fluctuates ; but when the stationary system is adopted, the whole
of the machinery must be employed to convey the goods, however
trifling.   Where the trade is great and nearly uniform, as is the
case between Liverpool and Manchester, the expense of the
stationary system approximates probably more nearly to that of
the locomotive than in any other locality in England.   It is in
this instance, therefore, that despatch and public accommodation
claim particular attention.

In this respect Mr. Walker is of opinion that either system is
fully adequate ; but he does not appear to have duly considered
the practical difficulties which are unavoidable where a chain of
stationary engines is employed.   Locomotive engines may be
compared to horses, as far as convenience is concerned, with
this advantage, that they are much more manageable, because
each engine is an independent power ; but the case is widely
different in the other system, where the whole is dependent on
each individual part, and also on a series of regulations liable
to be deranged by the inattention of workmen.   With the
locomotive engines the carelessness of one person extends in
most cases only to one train of carriages, whereas an accident

produced by the same cause with stationary engines occasions a
delay from one end of the road to the other, and the risk of
accident is evidently proportionate to the length of the line of
road.   We may go so far as to conceive a line of railway with
stationary engines so long that accidents would be almost per-
petually occurring, which leads to the inference that the conveyance of a large quantity of goods by such a series of engines
and ropes would, in the end, become *actually impracticable.*

From the local situation of the Liverpool and Manchester
Railway, it is evident that in a very few years several branches
from the various towns on each side will join the main line.
Hence the traffic on the different parts of the line will annually
undergo some modification.   If, however, the stationary system
were established in the outset, it would be necessary to construct
the engines of sufficient power to meet any probable increase of
the kind, and much difficulty would arise in adjusting the power
of the engines requisite for the different parts of the line of
road, excepting, indeed, they were all made very powerful to
meet any future increase of trade.   These and other difficulties,
which are inseparable from the stationary system on a public
line of road, where the trade must necessarily fluctuate, are
easily and effectually obviated with locomotive engines; for
should the trade in any part undergo a temporary increase, or
decrease, the requisite power may be immediately applied, or
withdrawn.

Many other practical considerations might be adduced to
exemplify the great superiority of the locomotive over the
stationary system on a public railway, but they are of a description not easily understood or duly appreciated, except by
those who have had experience and frequent opportunity of
witnessing the daily operation of machinery on railways.   Obstacles often arise from casualties, which by bare mention in
this place would appear frivolous; whereas to the practical man
they are of importance, and tend to demonstrate that it is of
great consequence to adopt a system, the efficient operation of
which, as a whole, is not dependent on each individual part.

The result of the locomotive competition on the
Liverpool and Manchester Railway gave such a strong

confirmation to the decision of the directors, by exhibiting
the great power of the locomotive, that it appears to have
put the rival system out of sight for some years; but the
idea of stationary power was too plausible not to be
revived.   The broad fact was obvious enough, that to
convey an engine and its boiler through the air, at a speed
of twenty or thirty miles an hour, exposed to all
weathers, and to all accidents of a precarious road, and
consuming a large portion of its energy in moving its
own weight along, was not the most advantageous way
of applying steam-power.   It was not wonderful, therefore, that any plan should be received with favour, which
offered an apparently feasible application, to railway
haulage, of the power of an engine, safely housed with
its various appurtenances in a fixed building, where it
could be worked in the most advantageous way.

The invention of the atmospheric plan of propulsion
offered this promise.   It was assumed that the principal
fault of the stationary system, as hitherto applied, lay in
the *rope*, as a means of transferring the power from the
fixed engine to the moving train; and undoubtedly,
by its great friction and liability to accident, the rope
was a serious evil.   The principle of the atmospheric
invention consisted simply in substituting for the rope
another means of transmitting the power, which was
conceived to be much superior, and the introduction of
which, it was thought, would entirely remove the objections to the stationary system, and leave its admitted
advantages available, without the drawbacks that had
formerly interfered with its application.

The first idea of transmitting power to a distance by

means of pneumatic pressure appears to belong to the celebrated Denys Papin, who described, in 1688, an apparatus, in which a partial vacuum, produced in a long tube by air-pumps fixed at one end, caused the motion of pistons placed at the other end. Mr. Farey, a well-known mechanical authority, writing of this scheme of Papin's in 1827, says : ' It is rather surprising that so simple and advantageous a method of exerting power at a distance from the first mover, should have remained neglected and unnoticed so long as it has been.'

But the more proximate inventor of the system of which we are now treating, was a London mechanical engineer named George Medhurst,* who, long before railways were thought of as a general means of convey-ance, proposed and described a plan of locomotion by atmospheric pressure, precisely similar in principle to that afterwards used.

His first notions on the subject were published in 1810, when he described his invention of ' New machinery for the rapid conveyance of letters, goods, and passengers by air.' He proposed to enclose letters and papers in a light hollow vessel, so formed as to fill the area of a tube, and to move freely through it; when, by forcing air into one end of the tube, he assumed that he would be able to drive the vessel through the tube at a great velocity.† He further proposed to extend the principle by making the tube large enough for a four-wheeled

---

* Medhurst has sometimes been described as a Danish engineer ; but the only explanation to be given for this statement is that he lived in Denmark Street, Soho !

† Exactly as has been lately pro-posed by the ' Pneumatic Convey-ance Company.'

carriage to run inside it, on an iron road, carrying goods and passengers through the kind of tunnel so formed.

In 1812 he again published a notice of his scheme, but adding the important conception that the necessity of putting the passengers and goods within the tube might be avoided, by substituting a much smaller pipe, the piston of which should communicate 'by a particular contrivance through the side of the tube' with the carriage outside, and so drag it along. The nature of this 'particular contrivance' he did not, however, disclose till 1827, when, in a third pamphlet, he described various modes of effecting this object. The mode with which we have to do here, introduced, on the top of the tube, a kind of longitudinal flap, riveted along one side, but loose on the other. The piston, running within the tube, had a wheel in front, which, as it passed along, lifted up the flap, forming a slit sufficient to allow a bent rod to pass through from the piston on the inside to the carriage on the outside, so as to give motion to the latter as the former was propelled by the pressure of the air. When the piston had passed, the flap closed of itself, the loose edge falling against a face of leather, or some other soft yielding substance, which made a joint sufficiently tight to prevent leakage under the small pressure the inventor proposed to employ.

Another important step in Medhurst's scheme of 1827 was that he proposed to work his piston both ways—one way by forcing air into the tube behind the piston, or by what may be called the *plenum* impulse;—the other way by exhausting the air before the piston, or by the *vacuum* impulse. Taking this latter modification, we have the perfect anticipation, as far as the idea is concerned, of the atmospheric system subsequently introduced upon railways. The merit of later inventions consisted in the perfection of the details of the apparatus, which Medhurst does not seem to have considered with much care. His invention was probably far too much in advance of the then notions of locomotion to meet with any encouragement for its actual trial.

It is right to mention that the first publication of any proposal for locomotion by Papin's *vacuum* principle appears to have been by a Mr. John Vallance of Brighton, who, in 1824, re-proposed Medhurst's plan of a gigantic tube, substituting, however, the vacuum for the plenum mode of action.

An agitation was got up at Brighton a year or two later for the trial of Vallance's plan between that town and London. A short trial-tube of the full size was constructed and worked by way of experiment; and it is well remembered how, for want of due precaution in checking the impetus of the carriage, the venturesome experimental passengers were occasionally blown out of the end of the pipe into a field beyond. But this attempt only had the result of furnishing jokes for the pantomimes of the day, and of producing a rather acrimonious paper war between the supporters and the opponents of the scheme. Mr. Vallance does not seem to have known or thought of the much more feasible plan of smaller tubes.

In 1834 Mr. Henry Pinkus, an American engineer, proposed a modified contrivance for opening and closing a slit at the top of the tube, by means of a flexible valvular cord. This was patented in the same year, as

also were other modifications for the same object in
1835 ; but the proposition appears to have had no practical result, and Medhurst's ideas remained in abeyance
until a few years afterwards, when Mr. Samuel Clegg, an
engineer well-known for the important part he took in
the introduction of gas-lighting, turned his attention to
the subject.   After studying it well, he adopted Medhurst's general arrangement of a vacuum tube, with a
longitudinal slit in its top, but he contrived a form of
valve for closing the aperture much superior to any
that had preceded it.   This was patented by Mr. Clegg,
January 3, 1839, and a tract giving an account of the
whole system of locomotion thus arranged, and calling
attention to its advantages, was published in the same
year.

But Mr. Clegg did not work alone in the matter, for,
before the date of his patent, he had associated himself
with an engineering firm—Messrs. Jacob and Joseph
d'Aguilar Samuda, eminent manufacturing engineers of
Southwark—who, apparently impressed with the value
of the invention, lent it powerful aid, not only by their
mechanical and engineering skill, but by the energy with
which they advocated its advantages.   Messrs. Samuda,
in 1844, obtained a supplementary patent for improvements, and through all the experiments and discussions
which took place on the subject they were the most active
and prominent supporters of the plan.

Soon after the date of the patent, Messrs. Clegg and
Samuda laid down at their own premises and elsewhere
small model tubes, which answered their expectations ;
but, not being satisfied with private experiments, they
endeavoured to get the system tried actually upon a

railway, and accordingly obtained permission to lay down, at their own expense, an experimental length upon a short line at Wormwood Scrubs, which had been made to connect the London and Birmingham and the Great Western Railways with the Kensington Canal, and which, though only a mile or two in length, was dignified with the name of the 'Birmingham, Bristol, and Thames Junction Railway.' A vacuum pipe, half a mile long and nine inches internal diameter, was laid down on the part of the line between the Great Western Railway and the Uxbridge Road, where the gradient was about 1 in 115, and where therefore the efficiency of the power in ascending inclines was put to the test.

This was set to work in June 1840; and as it was a complete exemplification, on a real scale, of the proposed plan of atmospheric propulsion, it may be as well to insert in this place, once for all, the description of the apparatus as given by the inventors.

The accompanying figures will serve to illustrate the description.

Fig. 1 is a general side view of the front part of the train and the atmospheric tube, the latter being delineated partly in longitudinal section to show the piston, and its attachment to the leading carriage.

Fig. 2 is a transverse section of the same parts.

Figs. 3, 4, and 5 are transverse sections of the tube only, enlarged to show the details more clearly, and to explain the action of the valve.

In fig. 3 the valve is shown open, the piston passing through; fig. 4 shows the method of closing the valve

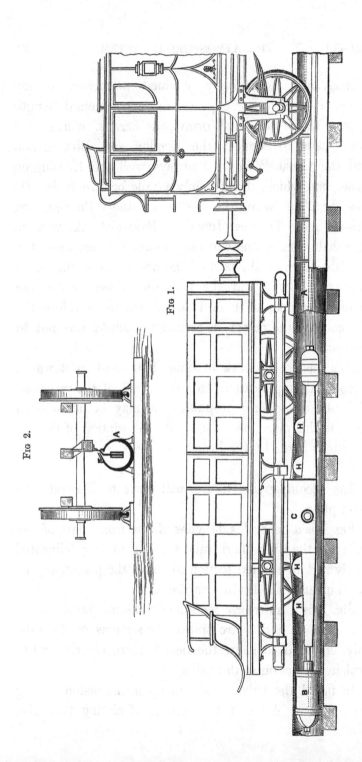

Fig 1.

Fig 2.

and sealing the composition; and Fig. 5 represents the valve as finally left after the carriage has passed by.

The same letters refer to the same parts in all the figures.

The moving power is communicated to the train by means of a continuous pipe or main A, laid between the rails, and divided by separating valves into suitable and convenient lengths for exhaustion. A partial vacuum is

Fig. 3.

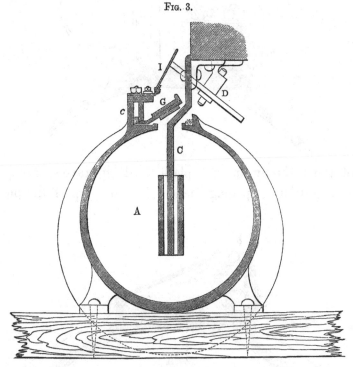

formed in each length of pipe by steam engines and air pumps fixed at intervals along the road. The separating valves are opened by the train as it advances, without stoppage or reduction of speed.

A piston B, which is made to fit air-tight by a leather

packing surrounded by tallow, is introduced into the
main pipe, and connected with the leading carriage

Fig. 4.

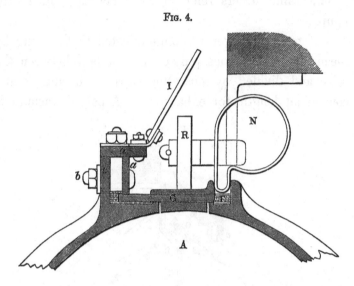

of the train by an iron plate c, which travels through
a longitudinal opening made along the top of the pipe

Fig. 5.

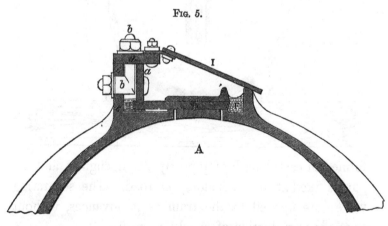

for its whole length. This opening is covered by a
valve G, extending also the whole length, formed of a

strip of leather riveted between iron plates; the top plates are wider than the groove, and serve to prevent the external air from forcing the leather into the pipe when the vacuum is formed; the lower plates fit the groove when the valve is shut, as shown in Figs. 4 and 5, and, by making up the circle of the pipe, prevent the air passing the piston. One edge of this valve is securely held down by iron bars aa fastened by screw bolts bb to a longitudinal rib c, cast on the pipe on one side of the opening; and the leather between the plates and the bar, being flexible, forms a hinge as in a common pump valve; the other edge of the valve falls on the surface of the pipe on the opposite side of the opening, thus forming one side of a trough F, as shown in Figs. 4 and 5. This trough is filled with a composition of bees'-wax and tallow, which is solid at ordinary temperatures, but softens when slightly heated. The composition, when so heated and pressed down, adheres to the edge of the valve, which forms one side of the trough, and to that part of the pipe which forms the other, and so makes an air-tight junction between them.

Supposing now the air to be exhausted from the part of the tube in front of the piston; the atmosphere having free access to the part behind it, will press upon it with a force proportional to its area and the degree of exhaustion; and the effect of this pressure will be to propel the piston along the tube, dragging with it the leading carriage to which it is attached, and the train coupled behind.

As the piston advances, the valve G must be raised to allow the connecting plate c to pass, and this is effected by four wheels H H H H, fixed to the piston rod behind the piston : the aperture thus formed serves also for the free admission of air to press on the back of the piston. When the wheels have passed by, the valve falls again by its own weight.

But by the operation of raising the valve out of the trough, the composition between it and the main pipe has been broken, and the air-tight contact must be reproduced. To effect this, another wheel R (Fig. 4) is attached to the carriage, which serves to ensure the perfect closing of the valve by running over the top plates immediately after the piston rod has passed ; and a copper tube or heater N, about five feet long, filled with burning charcoal, is also fixed to the under side of the carriage, and passes over the surface of the composition, softening it and pressing it down, so that when on cooling it becomes solid, it seals the joint air-tight as before. Thus each train, in passing, leaves the pipe in a fit state to receive the next train.

A protecting flap or cover I, formed of thin plates of iron about five feet long, hinged with leather, is made to lie over the valve, to preserve it from snow or rain ; the end of each plate underlaps the next in the direction of the piston's motion, being lifted up by wheels D (Fig. 3), fixed under the advancing carriage, and allowed to close again as it retires.

The parts above described constitute the essence of the plan. Much ingenuity and care were bestowed on the arrangement of other details, such as the entrance, exit, and separating valves, the mode of effecting junctions and crossings, the construction of the tube, the manner of connecting together the pipes of which it was formed, the arrangement of the exhausting pumps, &c. &c. But it is not necessary here to go into these particulars.

The exhausting pumps on the experimental line at Wormwood Scrubs were worked by a steam engine of fifteen horse-power, and produced in one minute a vacuum in the pipe equal to about 18 or 20 inches of mercury; and by maintaining this exhaustion, it was found that, even with the small pipe used, a load of $13\frac{1}{2}$ tons could be propelled up the incline at a rate of 20 miles an hour; or with a vacuum of $23\frac{1}{2}$ inches, a load of 5 tons would go 45 miles an hour.

These trials were considered so successful, and seemed to promise so much for the new system of propulsion, that they naturally attracted the attention of persons interested in railways, and among these was Mr. James Pim, treasurer of the railway between Dublin and Kingstown, who, after having carefully observed the experiments, became a most energetic advocate of the plan. It appears that the railway with which he was connected, had a short piece of line from Kingstown to Dalkey, which had been used for the transport of stone for the new works at Kingstown Harbour, and which, having steep gradients and sharp curves, offered what were then considered rather formidable difficulties to the working of the line.

About May 1841, Mr. Pim wrote a letter to Lord Morpeth, asking the permission of the Board of Public Works of Ireland (under whose care the Kingstown and Dalkey road was placed) for the parties interested in the experiment to lay down an atmospheric apparatus along this line.

It appears, however, that the sanction of the Board of Trade was needed to carry out this proposal, and Mr.

Pim, nothing daunted, wrote towards the end of the same year a letter to the Earl of Ripon, President of the Board, describing clearly the principle and mode of working of the atmospheric system, and giving a lucid and forcible statement of the arguments in its favour; the object of the letter being to ask the Board to refer it to such persons as their Lordships might select, to enquire into the several statements made, and in their report to state particularly whether the invention was entitled to a further and more extended trial.

The request was granted, and two scientific referees— Lieut. Col. Sir Frederic Smith, R.E., F.R.S., and Professor Barlow, F.R.S.—were accordingly appointed to investigate the merits of the plan.

Their report was dated February 15, 1842. It appears that they conducted experiments, in January, on the model line at Wormwood Scrubs, which generally corroborated those of the projectors; and after making the necessary calculations and deductions, they reported that they considered the principle of atmospheric propulsion to be established; that its economy of working would increase with the scale on which it was applied; and that it appeared well suited for such a line as that from Kingstown to Dalkey. On the points of first outlay, cost of working, safety, and convenience, as compared with the locomotive system, the referees did not venture any very decided opinion.

The report was, however, sufficiently favourable to warrant the Government in sanctioning the trial of the principle on the Kingstown and Dalkey line, and in granting a loan of £25,000 to the Company for the purpose. This determination was due to the influence of the late

Sir Robert Peel, then First Lord of the Treasury, who was on this occasion, as well as during its whole history, a strong supporter of the plan.

With this encouragement, the railway was accordingly prepared, the tubes laid down, the engines erected, and the apparatus was set to work in August 1843.

The line was single, and about one mile and three-quarters in length. It had a short descent from Kingstown, after which it rose to Dalkey with an average gradient of about 1 in 116; the steepest part, 1 in 57½. There was one considerable curve, of 518 feet radius; a shorter one of 570 feet radius; and a third, of 700 feet radius. The atmospheric tube was 15 inches internal diameter, placed in the middle of the road, between the two rails, and firmly attached to the cross transoms under the sleepers. It was in lengths of 9 feet each connected by socket joints carefully filled with cement. The width of the longitudinal valve opening was two and a half inches. The arrangements of the valve were made with all possible care, and with the benefit of all the experience gained by the previous experiment.

The pipe did not extend the whole length of the road, but stopped short of the summit of the hill by 560 yards, the carriages running up this distance by their momentum alone.

The steam engine was placed, for the sake of convenience, at about 500 yards from the upper end of the tube, being joined to it by a connecting pipe of equal diameter. The engine was of 100 horse-power, working an air pump of 67 inches diameter, with a stroke of 5 feet 6 inches.

Experiments made on the line, soon after the opening,

gave good results as to the action of the apparatus. It
was found that a rarefaction of 13 to 14 inches could be
obtained in two minutes, and 22 inches in five minutes;
the pump making 22 strokes per minute. And in running
trains up the incline, it was observed that 30 tons could
be drawn up at a speed of about 30 miles an hour, and
70 tons at about 20 miles; which, considering the difficulties of the road, was certainly satisfactory.

This confirmation of the results previously obtained
on a smaller scale served to increase the popularity
of the new invention, and to stimulate its promoters to
urge its claims upon the railway world, with a view to
securing its more extended adoption.

It may be well here to give a summary of the chief
arguments which, at various periods, were urged in favour
of the atmospheric system, as compared with other modes
of railway locomotion.

The more cogent of these took the shape of objections
to the locomotive engine. It was said—

(1) In the first place, that to make a steam engine
locomotive was eminently unfavourable for its economy
of fuel; that the quantity consumed was excessive, and
the kind expensive.

(2) That this was also a very unfavourable condition
for keeping the engine in repair, and that the necessity of
having a large stock of engines constantly under examination in 'hospital' led to a very large extra outlay of
capital.

(3) That the locomotive engine had to overcome the
friction and other resistances due to itself, and to the tender
carrying its supplies of fuel and water; to which had

also to be added a resistance peculiar to this machine, that of the back pressure on the pistons caused by the blast-pipe.

(4) That in addition to this loss, it had also, in ascending gradients, to overcome the gravity of itself and its tender.

(5) That the use of the locomotive involved many other minor evils—such as the necessity for repairing shops and running sheds, distributed over the line; the liability to slipping on the rails, to fire, to bursting, to freezing of the pumps, and to many accidental causes of derangement and mischief which did not exist with stationary power.

Such were the principal evils said to be inherent in the locomotive system. The only form of stationary power with which the atmospheric plan could be compared, was that of the rope, and to this it was objected —

(6) That the friction of the rope was enormous, and that in ascending inclines the weight of the rope was also the source of much loss of power.

The advantages peculiar to the atmospheric system were stated to be:—

(7) That it got rid of all the disadvantages named in the first five heads, as inherent in the travelling form of the motive machine, and was free also from the objections to the use of a rope with stationary power, as the air in the tube did the duty of the rope without either weight or material friction.

(8) That it presented much greater safety than the locomotive plan, for several reasons,— that it was quite impossible any two trains could come into collision, either by meeting or overtaking each other; that the

leading carriages could not get off the rails; and that all the manifold elements of danger inherent in the locomotive were avoided.

(9) That any desired speed of travelling might be obtained, by simply proportioning the engine, pumps, and pipe accordingly, without corresponding disadvantage in the application of the power. And that, therefore, higher speeds might be attained on railways generally.

(10) That as a consequence of the more favourable application of the power, and the less danger of getting off the line, much steeper gradients and sharper curves might be used, than on lines prepared for locomotive haulage; and that consequently the cost of constructing railways might be very much lessened; the economy being further enhanced by the reduced height of all tunnels and over-bridges, consequent on the absence of the locomotive chimney—and the less strength required for viaducts and under bridges, which would have less weight to carry.

(11) That a further and still greater saving in first cost would result from the fact, that the principle of atmospheric propulsion, by ensuring regularity in the working of the trains, would admit of a single line being used, with safety, for an amount of traffic which on the locomotive system must imperatively demand a double line.

(12) That by doing away with the heavy locomotive, much might also be saved in the first cost and in the maintenance of the permanent way; as lighter rails might be used, and they would be much less liable to deterioration and derangement.

(13) It was further contended that the atmospheric system offered much more convenience to the public,

inasmuch as it would be the interest of the companies, under this system, to despatch light trains very frequently; whereas the use of locomotive power rendered it advantageous to reduce their number, and concentrate their weight, as much as possible. And it was also added that the atmospheric system was much more agreeable to the passengers, for several reasons, such as the entire absence of dust and sparks from the engine, less noise, more steady and comfortable motion, better condition of the road, &c. &c.

(14) And finally, it was said that the atmospheric system would enable water power to be used, where it existed, instead of steam; and that, where a sufficient quantity and fall could be obtained to produce a vacuum, machinery might be dispensed with altogether.

The popular and plausible nature of many of these arguments could not fail to attract the attention of the public; particularly as the new plan proposed was no mere untried scheme; for it was in actual practical application, working the traffic daily over a line which, for locomotives, at that time, had been admitted to be almost impracticable.

It was no wonder, then, that the atmospheric system, working on the Kingstown and Dalkey line, at the end of the year 1843, should be carefully examined by railway engineers; and among the first to give attention to it was Robert Stephenson.

About this time an application was made to the Directors of the Chester and Holyhead Railway, (who were then promoting their Bill in Parliament,) with a view to the application of the system on that line. The directors, feeling that an investigation ought to be made, commissioned Mr. Stephenson, their engineer, to examine the invention, and to report to them whether he could recommend its application to their railway.   He undertook two series of careful investigations, and his report to the directors thereon was dated April 9, 1844, only a few months after the opening of the Dalkey line; so that Mr. Stephenson appears to have been the first independent investigator of the system, in its application on a practical scale.   The results of this scrutiny are so important, when taken in connection with what afterwards occurred, that it is necessary to give them at some length.

Mr. Stephenson commences his report by a passage which well illustrates the importance he attached to the investigation.   He says to the Directors :—

When I first visited Kingstown at your request, I made such experiments as appeared sufficient to enable me to form an accurate opinion on the application of this new motive power to public railways.   On my return to England, however, I found, by analysing the experiments, that many of the results were irreconcilable with each other, presenting anomalies in themselves, and suggesting further enquiry.

It was then that I began to feel the onerous and difficult nature of the task I had undertaken.   I was called upon, in short, to decide whether a singularly ingenious and highly meritorious invention was, or was not, to be applied to the Chester and Holyhead Railway.   I also felt strongly that whatever might be my opinion, whether favourable or unfavourable, the final destiny of the invention was not in my hands; and that if it were really calculated to produce the remarkable results which had been stated, nothing could stop its universal application to railways.   On the other hand I saw that, if the principles of the invention were not soundly based, I should be incurring a most serious responsibility in recommending its application to the Chester and Holyhead Railway, extending over a distance of eighty-five miles.

Under this conviction, I arranged an entirely new and extended series of experiments, with the view of fully and accurately testing every part of the invention, and thus putting myself in a position to give you an opinion upon which I could recommend you to act.

Mr. Stephenson further paid a deserved compliment to the engineers who had introduced the system, by stating that 'the mechanical details of the apparatus employed at Kingstown had been brought to a remarkable degree of perfection.'

In commencing his investigations Mr. Stephenson first took means to test the actual capabilities of the apparatus, irrespective of any hypothesis, by ascertaining the maximum velocity attainable with trains of various weights, noting also the corresponding pressures in the vacuum tube; and an elaborate statement is given of 20 experiments of trains actually conveyed up the incline, gradually increasing in weight from $23\frac{1}{4}$ to $64\frac{3}{4}$ tons. The general results may be thus stated.

With the lightest trains, of 23 to 25 tons, a velocity of 30 to 35 miles an hour was attained, with a vacuum of 13 to 17 inches of mercury.

With medium trains of 40 to 45 tons, a velocity of about 25 miles an hour was arrived at, with 22 inches of vacuum.

With the heaviest trains of 60 to 65 tons, a speed of 16 to 18 miles an hour was attained, with a vacuum of $23\frac{1}{2}$ to $24\frac{1}{2}$ inches.

Mr. Stephenson proceeded to reason upon these actual facts exhibited. He showed that, supposing the apparatus to be in every respect perfect, the velocity of the piston in the tube, when uniform motion was attained, would be, to that of the air-pump piston, inversely as their areas;

but that, from various imperfections inherent in the system, this was not practically the case.

The nature and influence of these imperfections therefore formed the next subject of investigation, and the principal of these was the leakage of air in consequence of various defects in the joints, but principally through the longitudinal valve at the top of the tube. He tried a series of experiments to determine this, and came to the conclusion that whatever might be the degree of rarefaction of the tube, nearly equal volumes of air, measured at atmospheric pressure, would leak into the tube in equal times; this curious result being apparently due to the fact that at high pressure (when greater quantities might have been expected to enter) the valve was forced closer, and the apertures of leakage were reduced in proportion. The average amount of leakage, measured at atmospheric pressure, he found to be 186 cubic feet per minute per mile of tube, or 252 feet for the whole length, to which had to be added 219 cubic feet per minute for the connecting pipe and air pump.

But Mr. Stephenson went on to show that, although the atmospheric volume leaking in was pretty uniform at all pressures, the effect of this, as regarded the power required to remove it from the tube, was extremely variable under different degrees of rarefaction; for since the entering air would become expanded according to the rarefaction, and since the air pump could only extract a fixed volume of the rarefied air at each stroke, the power and time required to overcome the effect of the leakage must increase very rapidly with the degree of exhaustion used. And hence, as the exhaustion advanced, the retarding influence of the leakage on the speed became more and more serious, and the maximum velocity attainable by the train proportionably lowered.

Having determined the value and effect of the leakage, Mr. Stephenson calculated what velocity the Dalkey tube ought to give, at the assumed ordinary speed of the air pump — first, supposing the apparatus to be perfect, and secondly, allowing for the effect of the leakage; the difference between which was found, with a vacuum ot 18 inches, to be 13 per cent, and with $24\frac{1}{2}$ inches to be 30 per cent, this difference expressing the calculated loss due to the leakage.

These calculations were then further tested by actual results of experiments with the trains, which showed that the real velocity attained fell still short of this latter result by quantities varying from 26 to 41 per cent, giving the total departure from the theoretical state of perfection 39 to 71 per cent.

The causes of this latter or additional loss of effect Mr. Stephenson attributed partly to further imperfections in the air pump, when in motion, beyond those observable when at rest; and partly to the leakage round the propelling piston, which he considered was much augmented during its swift motion in the tube.

Next followed a series of calculations on the power consumed in giving motion to the trains, under the various circumstances. These calculations were exceedingly elaborate, and, from the evident desire of Mr. Stephenson to present to his readers all the data which had led him to his conclusions, were made somewhat complicated and abstruse as well as lengthy. He had indicator diagrams taken from the air pump at various states of the rarefaction (which are published in full in the report), and by this means—taking, as before, the assumed speed of the air pump—he arrived at the power expended by the steam engines to produce the results obtained, some idea of which may be formed from the following statement:—

With a train of $26\frac{1}{2}$ tons, which attained a uniform velocity of 34·7 miles per hour, the total power expended was 322 horses.

With 45·5 tons, attaining 25·2 miles an hour, the power was 427 horses.

With 64·7 tons, at 16·7 miles an hour, it was 415 horses.

These amounts, however, included the power expended to raise the vacuum, and to start the train, which was generally more than what was necessary to keep it in uniform motion. The value of the latter came out practically at a pretty nearly uniform value of about 170 to 180 horses' power for all trains and speeds, and all degrees of exhaustion.

It was next calculated what portion of this was actually applied, through the tube piston, to the propulsion of the trains, which was found to be:—

For the $26\frac{1}{2}$ ton train, 150 horse-power; for the $45\frac{1}{2}$ ton train, 134 horse-power; for the 64 ton train, 96 horse-power: the loss (due to leakage) increasing with the degree of exhaustion applied.

Finally he estimated the component parts of the resistance offered to the motion of the trains, and after making the proper allowance for gravity (1 in 115) and friction (10 lbs. per ton), a large surplus was found to be due to the resistance of the atmosphere. For example, with the $26\frac{1}{2}$ ton train, moving at 34·7 miles an hour, this residual resistance was found to be 78 horse-power out of 150. With the 45½ ton train, at 25 miles an hour, it was 44 out of 134. And with the 64 ton train, at 16¾ miles, it was 11 horse-power out of 96.

Mr. Stephenson considered this last result as of great importance, and having a bearing much wider than the case in question. He says : —

In referring to the loss of power from the resistance of the atmosphere, it will be observed there is a very rapid reduction in the loss, as the speed is diminished, indicating most satisfactorily the excessive expenditure of power, and consequent augmentation of expense, in working at high velocities upon railways. This remark is of course equally applicable to all railways, whatever be the motive power employed, and it is here introduced only for the purpose of showing that the attainment of speed exceeding that which is now realised upon some of the existing lines of railway is a matter of extreme difficulty, and that the atmospheric system is not exempt from that wasteful application of power which high velocities inevitably entail. For although the resistance of the atmosphere to railway trains has been established for a long time, the limit which it is likely to put to every effort to obtain such velocities as have been generally believed to be within the reach of the atmospheric railway has not, I am sure, been sufficiently brought forward.

Mr. Stephenson also pointed out, as a result of his experiments, the necessity of working with only a moderate degree of exhaustion; as he was led to the conclusion that when the barometer rose to a certain height, the expansion of the air leaking into the apparatus must become fully equal to the total capacity of the pump, and no advance of the tube piston could be effected ; this case occurred on the Kingstown and Dalkey Railway, with a height of barometer of 25½ inches. 'This conclusion,' adds Mr. Stephenson, 'which is unquestionably correct, points out the improvident expenditure of power when a high degree of rarefaction is required.'

Having thus explained the object and result of the experiments instituted on the Kingstown and Dalkey Railway, Mr. Stephenson proceeded to draw a comparison between the working of the atmospheric system and of other descriptions of motive power, with the view of showing their relative advantages and disadvantages.

The first comparison was with the stationary engine and rope.   Mr. Stephenson chose the incline on the North-Western Railway, from Euston Square to Camden Town, nearly a mile long, and with an average gradient of 1 in 106, and which at that time was worked in this manner.   He gave a table of experiments upon it, and showed, by an example serving for comparison in the two cases, that the waste of power on the Euston incline amounted to only 45 per cent, as against 74 per cent on the atmospheric plan.   At the same time it was admitted that working a longer length of line would make the comparison more favourable to the latter plan.

Next came the comparison with the locomotive engine.   Mr. Stephenson took as an example the atmospheric train of $26\frac{1}{2}$ tons, moving at 34·7 miles per hour, at which rate he found the total loss of power by leakage, getting up the vacuum, and starting the train, equal to 53 per cent of the quantity developed by the engine.   He then found that, when a locomotive drew a train of the same weight up the same gradient, its own gravity, friction, and atmospheric resistance, ' together with a further resistance arising from the pressure of the atmosphere against the pistons, peculiar to the working of a locomotive,' would consume to waste

54 per cent of the total power developed. 'Therefore,' infers Mr. Stephenson, 'the loss of power by the use of the locomotive engine under such circumstances appears somewhat to exceed that shown by the atmospheric system; this is, however, a most disadvantageous comparison for the locomotive engine, because the *gradient far exceeds that upon which it can be worked economically.*'

'Such a comparison,' he says in another place, 'cannot be held as strictly correct, because the locomotive engine, as a motive power on steep gradients, is *wasteful, expensive,* and *uncertain*; therefore on a long series of bad gradients, extending over several miles, where the kind of traffic is such that it is essential to avoid intermediate stoppages, the atmospheric system would be the most expedient. The lightest trains taken upon the Kingstown and Dalkey incline, at the velocities recorded, probably exceed the capabilities of locomotive engines, and so far prove that the atmospheric system is capable of being applied to somewhat steeper gradients, and that on such gradients a greater speed may be maintained than with locomotive engines.'

As regarded lines of more moderate steepness, he reduced the Dalkey performance to what it might be held equivalent to on a level, in order to show that such a performance was exceeded on many locomotive lines; and he added the strong expression of his opinion that on lines of railways where moderate gradients were attainable at a reasonable expense, the locomotive engine was decidedly superior, both as regarded power and speed, to any results developed or likely to be developed by the atmospheric system.

Up to this point, the calculations and remarks in the Report had reference solely to the question of power,

entirely independent of the questions of expense or convenience, which therefore Mr. Stephenson next proceeded
to examine, beginning with the cost of construction.

The advocates of the atmospheric system, knowing the
great expense of the apparatus, had asserted it to be
possible to work any reasonable amount of traffic with a
single line.   This assertion Mr. Stephenson disputed,
showing that on a long line, if trains were despatched with
sufficient frequency to carry the traffic both ways, the
delay, by stoppages necessary for the trains to pass each
other, would be so great as to defeat the object altogether.
Hence he considered a double line absolutely essential for
any considerable length of railway; and he also concluded that each line must be provided with its proper
complement of engine-power.   'The intersections of the
trains,' he says, 'cannot possibly be made to take place
always at the same points, even on the supposition that
each railway is worked independently of every other
with which it may be in connection.   When we introduce, in addition, the fact that several branch lines must
necessarily flow into the main trunks; that no line can be
worked independently; that the arrival of trains is, and
must always be, subject to much irregularity, sometimes
arising from their local arrangements, sometimes from
weather, and at others from contingencies inseparable
from so complicated a machine as a railway; it must be
palpable that two independent series of stationary engines
are as indispensable as two independent lines of vacuum
tube, for the accomplishment of that certainty, regularity,
and despatch which already characterise ordinary railway
operations.'

Coming to figures, Mr. Samuda, on behalf of the

atmospheric system, had estimated for a single line of tube as follows :—

	Per Mile.
Vacuum Tube, with all its appliances . .	£3,342
Engines	1,343
	£4,685

Mr. Stephenson, considering a double line necessary, and that Mr. Samuda had not taken his engines powerful enough, altered this to—

	Per Mile.
Vacuum Tubes	£7,000
Engines	4,000
	£11,000

He then applied this, as an example, to the London and Birmingham Railway, 111 miles long, which would make the cost of the atmospheric apparatus amount to £1,221,000, whereas the capital expended on locomotives and all their contingent outlay was only £321,000, making a difference of £900,000 against the atmospheric system.

Mr. Stephenson admitted, that if that line had been originally laid out for the atmospheric plan, a saving of £900,000 might have been accomplished in the original design ; but he remarked that on other lines of railway where the gradients conformed more to the surface of the country, the excess of first outlay to adapt the atmospheric plan to them would be very heavy ; and he gave an example of a cheap railway for light traffic in Norfolk, where the application of the atmospheric system would have involved a cost so great as to render it totally inapplicable.

The next point considered was the cost of working, which, taking the London and Birmingham again as an example, Mr. Stephenson was of opinion would be greater by the atmospheric than the locomotive system in the proportion of £74,000 to £64,000 per annum.

Such were Mr. Stephenson's conclusions as to the capabilities of the atmospheric system as a motive power, and the cost of applying it. He finally devoted a short space to the consideration of some other questions, scarcely of less consequence when the application of the system to daily practical purposes was discussed; namely, the speed attainable, the safety, the certainty, and the liability to casualty or derangement.

These were questions, he said, upon which widely different opinions might be entertained, but some of which could only be fairly appreciated by persons really conversant with the practical working of railway traffic.

As to the *speed*, he considered he had already proved that, though increased velocity might be obtained, it could only be done with an inordinate expenditure of power.

On the *safety* of the atmospheric system there could, he said, be little room for difference of opinion, as it might be stated to be nearly perfect; but he thought that further experience would much diminish the risk with locomotive engines.

But the question of certainty of action, Mr. Stephenson conceived, would be found to involve considerations militating most seriously against the plan, even though the first outlay and cost of working were in its favour.

'Each train,' he says, 'in moving between London and Birmingham, would be passed, as it were, through thirty-eight distinct systems of mechanism, and it cannot be deemed unreasonable to suppose that in such a vast series of machinery as would be required in this instance, casualties occasioning delay must not unfrequently occur. If the consequences were confined to one train, such casualties would be of small moment, but the perfect operation of the whole is dependent on each individual part, and when the casualties extend themselves not only throughout the whole line of railway, but to every succeeding train which has to pass the locality of the mishap, until it is rectified, whether this occupies one hour or one week, the chances of irregularity must be admitted to be very great. The delay would apply to every train, whatever might be its destination, and to every railway in connection with that upon which the accident occurred. Such a dependency of one line of railway upon the perfectly uniform and efficient operation of a complicated series of machinery on every other with which it is connected, appears to me to present a most formidable difficulty to the application of the system to great public lines of railway; so formidable, indeed, that I doubt much whether, if in every respect the system were superior to that of locomotive engines, it could be carried out upon such a chain of railways as exists between London and Liverpool, or London and York.

'This difficulty, which is insurmountable and inherent in all systems involving the use of stationary engines, was fully considered previous to the opening of the Liverpool and Manchester Railway, when the application, to that line, of stationary engines and ropes was contemplated; at that time the objection of the whole line being so dependent upon a part was maturely weighed, and decided to be most objectionable. In going through this investigation, I have again deliberated much on the feasibility of working such a system, but without any success in removing those obstacles which must interfere with the accomplishment of that certainty which has become indispensable in railway communication.'

Mr. Stephenson also referred to other objections; such as the chance of derangement of the tube by subsidence of the earthwork; the complication of working the traffic at intermediate stations; the difficulty of shunting, or of stopping the train on a sudden emergency, &c. &c. Of these, with many other objections of a minor character, he chose to take only a slight notice, as his wish was to call attention only to the main features of the invention, and to treat nothing as a difficulty which was not obviously inherent or irremediable in the atmospheric system itself.

Finally, Mr. Stephenson summed up the conclusions to which his investigation had led him in the following terms :—

I. That the atmospheric system is not an economical mode of transmitting power, and inferior in this respect both to locomotive engines and stationary engines with ropes.

II. That it is not calculated practically to acquire and maintain higher velocities than are comprised in the present working of locomotive engines.

III. That it would not in the majority of instances produce economy in the original construction of railways, and in many would most materially augment their cost.

IV. That on some short railways where the traffic is large, admitting of trains of moderate weight, but requiring high velocities and frequent departures, and where the face of the country is such as to preclude the use of gradients suitable for locomotive engines, the atmospheric system would prove the most eligible.

V. That on short lines of railway, say four or five miles in length, in the vicinity of large towns, where frequent and rapid communication is required between the termini alone, the atmospheric system might be advantageously applied.

VI. That on short lines, such as the Blackwall Railway, where the traffic is chiefly derived from intermediate points, requiring frequent stoppages between the termini, the atmospheric system is inapplicable, being much inferior to the plan of disconnecting the carriages from a rope for the accommodation of the intermediate traffic.

VII. That on long lines of railway the requisites of a large traffic cannot be attained by so inflexible a system as the atmospheric, in which the efficient operation of the whole depends so completely upon the perfect performance of each individual section of the machinery.

Appended to Mr. Stephenson's Report was a statement by Mr. G. P. Bidder regarding the Blackwall Railway. This line was at that time worked by stationary engines and ropes, but the advocates of the atmospheric system had urged its adoption in preference.

Mr. Bidder, after describing the peculiar circumstances of the traffic on the line, showed satisfactorily that the atmospheric system could not be applied to it with advantage, chiefly from the necessity for frequent stoppages at intermediate stations, which could not be effected, or at least, if effected, would entail such delays as would be extremely inconvenient to the public, and prejudicial to the interests of the line.

Since 1849, the Blackwall Railway, having been extended and connected with other lines, has been worked by locomotive power.

Mr. Stephenson's Report was published and widely circulated; but though it decided the Chester and Holyhead directors not to adopt the atmospheric system on their line, it does not seem to have checked its advance in public estimation; for the features of the scheme were so attractive and popular as to secure for it the evident favour, not only of railway authorities and the public generally, but also of a number of professional engineers. To the latter class the subject naturally proved a very interesting one; for at three several times, in the years 1844 and 1845, it was brought prominently forward and discussed at great length at meetings of the Institution of Civil Engineers, almost all the principal members of the profession taking part in the arguments either on one side or the other; and at

a later period, when it was in action on the Croydon
line, the same Institution promoted the appointment of
a scientific committee to make experiments upon it—
a design which, had it not been frustrated by the sudden
and premature abandonment of the system on that line,
would undoubtedly have been of the greatest interest and
importance in a scientific point of view.

Almost immediately after the date of Mr. Stephenson's
Report, the subject was brought forward in Parliament, by
a bill promoted by the Croydon Railway Company for an
extension of their line to Epsom, which it was proposed to
work on the atmospheric plan.   On the committal of this
bill a long investigation into the merits of the system took
place, extending from the 15th to the 21st of May, and
embracing all that could be said for or against the plan.
Mr. Cubitt, the engineer of the line, explained his reasons
for adopting it, in which he was supported by Mr. Samuda, Mr. Gibbons of the Dalkey line, and Mr. I. K. Brunel.
Mr. Stephenson gave evidence against it, stating and explaining the arguments used in his report to the Chester
and Holyhead Railway.   The Committee, however, appear
to have been satisfied of the practicability of the plan, as
they passed the bill ; and when it had gone through the
other legal stages, steps were immediately taken to put the
works into execution.

In the next year, 1845, the subject attracted still more
prominently the public attention.   This, it will be recollected, was about the time of the well-known railway
mania, when speculation was excited to a degree unheard
of before, and new lines were promoted in vast numbers
from all corners of the empire.   The atmospheric
system, by promising to cheapen the construction of new

lines, and to facilitate their formation, was too enticing to be overlooked; and consequently, among the multitudes of new lines introduced into Parliament at the commencement of the session of 1845, were many in which this system of propulsion was proposed to be adopted, and some of which indeed, by their inapplicability to locomotive traction, depended for their very existence on the feasibility of the plan.

Independently of the Epsom line, already sanctioned and in progress of construction, many others, of much public importance, were now proposed to be worked on this system—as, for example, one from Newcastle to Berwick ; continuations of the Epsom line directly to Portsmouth, of the Croydon line to Maidstone, Tunbridge, and Ashford, and of the Dalkey line to Bray, and a direct line from London to Northampton.

To have left the full discussion of the atmospheric principle to be undertaken in each separate case would, it was thought, lead to much difficulty and delay ; and therefore, on the motion of Lord Howick (afterwards Earl Grey), a Committee of the House of Commons was appointed to investigate, once for all, the merits of the plan, and to report to the House thereon.

The Committee consisted of fifteen members, the Hon. Bingham Baring in the chair. They were appointed March 14, 1845, examined witnesses from April 1 to April 11, and made their Report on April 22 ; an instance of very remarkable expedition, showing the urgency they attached to the subject.

The first witness examined was Mr. Samuda, one of the patentees, who explained at great length the nature and advantages of the principle; stated what was being

done to put it into operation on various lines; and
answered objections that had been made against it.

Mr. Barry Gibbons, the engineer, and Mr. Bergin, the
manager, of the Kingstown and Dalkey line, explained the
working of the system there, and testified to its success.

Mr. Brunel supported the plan, and described the
extensive use he was making of it on the South Devon
line. Mr. (afterwards Sir) William Cubitt followed on
the same side, and described his application of the atmospheric system to the Croydon and Epsom lines. The
principle was also supported by Mr. Vignoles and Mr.
Field, eminent civil and mechanical engineers, and by the
Rev. Dr. Robinson of Armagh.

On the other hand, Mr. Robert Stephenson, Mr. Bidder,
Mr. Nicholson, and Mr. Locke testified against the system.

The advantages of atmospheric propulsion, and the arguments in its favour, which have been already stated, were
urged upon the Committee by the first set of witnesses,
while the objections to it were principally those given in
Mr. Stephenson's report, to which, however, a few minor
ones were added—as the loss of power by the friction
of the air in the tube; the heating of the air in the
pump during compression; the impossibility of making
level crossings, or of having junctions with branch lines
except at the principal stations; the great cost of
running only few trains, as at night, and so on.

The Committee, after due deliberation and discussion
among themselves, adopted a Report which is of sufficient
importance to warrant its insertion here. It runs as
follows, omitting some passages of minor interest :

The Select Committee appointed to inquire into the merits of
the atmospheric system of railway have examined the matters
to them referred, and have agreed to the following Report.

Your Committee have given their best attention to this interesting subject. Adverting to the great number of Railway Bills now in progress, they consider that one of the most practical results of this inquiry would be lost if their Report were delayed until after these bills had passed through Committee, and a decision had already been made on their comparative merits.

Your Committee have endeavoured therefore to present to the House, with as little delay as is consistent with the due discharge of their duty, the evidence which they have taken, and the opinions to which they have come, and they trust that their labour may not prove altogether useless to the Committees that have to decide on the particular railway schemes now pending.

The House are aware that a railway on the atmospheric principle is already in operation between Kingstown and Dalkey, in Ireland.

The first object of your Committee was to make a full inquiry into the result of this experiment. From Mr. Gibbons, Mr. Bergin, and Mr. Vignoles, gentlemen officially connected with the Kingstown and Dublin, and Kingstown and Dalkey Railways, they received the fullest and frankest evidence on all the points connected with their management

From this evidence, and from that of Mr. Samuda, it appears that the Dalkey line has been open for nineteen months, that it has worked with regularity and safety throughout all the vicissitudes of temperature, and that the few interruptions which have occurred have arisen rather from the inexperience of the attendants than from any material defect of the system.

Your Committee find, moreover, that high velocities have been obtained with proportional loads on an incline averaging 1 in 115, within a course in which the power is applied only during one mile and an eighth.

These results have been displayed under circumstances which afford no fair criterion of what may be expected elsewhere; for in addition to the curves on the line, which would have been considered objectionable if not impracticable for locomotive engines, there are alleged to exist defects in the machinery and apparatus, occasioned partly by the difficulties of the situation, partly by mistakes inseparable from a first attempt, which very seriously detract from the efficiency of the power employed, for the remedy of which provision has been made in the experiments now in progress.

These are important facts. They establish the mechanical efficiency of the atmospheric power to convey with regularity, speed, and security, the traffic upon one section of pipe between two termini; and your Committee have since been satisfied, by the evidence of Messrs. Brunel, Cubitt, and Vignoles, that there is no mechanical difficulty which will oppose the working of the same system upon a line of any length. They are further confirmed in this opinion by the conduct of the Dalkey and Kingstown directors, who have at this moment before Parliament a proposition to extend their atmospheric line to Bray.

In addition to the witnesses already mentioned, your Committee have had the advantage of hearing the objections urged by Messrs. Nicholson, Stephenson, and Locke against the adoption of the atmospheric principle, and the grounds of their preference for the locomotive now in use.

Your Committee must refer the House to the valuable evidence given by these gentlemen.

It will be seen that great difference of opinion exists between them and the other witnesses to whom your Committee have before referred, both in their estimation of what has already been effected, and in their calculations of future improvement.

But without entering upon all the controverted points, your Committee have no hesitation in stating that a single atmospheric line is superior to a double locomotive line in both regularity and safety, inasmuch as it makes collisions impossible, except at crossing places, and excludes all the danger and irregularity arising from casualties to engines or their tenders.

Your Committee desire also to bring to the attention of the House a peculiarity of the atmospheric system, which has been adduced by the objectors to prove how unsuited it must be profitably to carry on a small and irregular traffic; namely, that the greatest proportion of the expenses of haulage on the atmospheric principle are constant, and cannot be materially reduced, however small the amount of traffic may be. This is, no doubt, a serious objection to the economy of the atmospheric system under the circumstances above alluded to. But on the other hand, as the expenses do not increase in proportion to the frequency of the trains, it is to the interest of companies adopting the atmospheric system to increase the amount of their traffic by running frequent light trains at low rates of fare, by which the convenience of the public must be greatly promoted.

Upon an atmospheric railway the moving power is most economically applied by dividing the weight to be carried into a considerable number of light trains. By locomotive engines, on the contrary, the power is most conveniently applied by concentrating the traffic in a smaller number of heavier trains. The rate of speed at which trains of moderate weight can be conveyed on an atmospheric line makes comparatively little difference in the cost of conveyance, while the cost of moving trains by locomotive engines increases rapidly with the speed.

Now when it is considered that we surrender to great monopolies the regulation of all the arteries of communication throughout the kingdom, that it depends in a great measure upon their view of their interest when we shall travel, at what speed we shall travel, and what we shall pay, it becomes a material consideration, in balancing the advantages ensured to the public by rival systems, to estimate not so much what they respectively can do, but what, in the pursuit of their own emolument, they will do.

The main objections of the opponents of the atmospheric system seem to rest—first, on the supposed increase of expense of the atmospheric apparatus over and above the saving made in the construction of the road ; secondly, on the inconvenience and irregularity attending upon a single line. With reference to the last point, your Committee felt it their duty to direct their first attention to the question of security, and they have already stated that there is more security in a single atmospheric line than a double locomotive. They may further observe that they find the majority of the engineers who have been examined are decidedly of opinion that any ordinary traffic might be carried on with regularity and convenience by a single atmospheric line.

With respect to expense, and to some other contested points, your Committee do not feel themselves competent to report a decided opinion. It would scarcely be possible at the present time to institute a fair comparison of a system which has had fifteen years of growth and developement, with another which is as yet in its infancy. That comparison would, after all, be very uncertain ; it must depend much on details of which we are ignorant, much on scientific knowledge which we do not possess. There are, however, questions of practical importance, having reference to the present state of the Railway Bills before the House, to which your Committee consider themselves bound to advert.

There is a doubt, raised in the Reports of the Board of Trade, whether the atmospheric system has been sufficiently tested to justify the preference of a line which can only be worked on the atmospheric system, or which presents gradients less favourable than a competing line for the use of the locomotive engine.

If it were practicable to suspend all railway legislation until the result of the Devon and Cornwall, and of the Epsom and Croydon atmospheric lines were known, it would be perhaps the most cautious and prudent course to wait that result; but such a course, independent of all considerations of expediency, is evidently impracticable. Your Committee venture therefore to express their opinion to the House, that in deciding between competing lines of railway, those which have been set out to suit the atmospheric principle ought not to be considered as open to valid objections merely on account of their having gradients too severe for the locomotive; nor should they be tested, in comparison with other lines, solely by the degree of their suitableness to the use of the locomotive.

No doubt, in matters like these, experience alone can decide the ultimate result; but your Committee think that there is ample evidence which would justify the adoption of an atmospheric line at the present time. All the witnesses they have examined concur in its mechanical success.

Mr. Bidder says: ' I consider the mechanical problem as solved, whether the atmospheric could be made an efficient tractive agent. There can be no question about that, and the apparatus worked, as far as I observed it, very well. The only question in my mind was as to the commercial application of it.' Mr. Stephenson admits that under certain circumstances of gradients (1,315), and under certain circumstances of traffic, without reference to gradients (1,204), the atmospheric system would be preferable.

While your Committee have thus expressed a strong opinion in favour of the general merits of the atmospheric principle, they feel that experience can alone determine under what circumstances of traffic, or of country, the preference to either system should be given.

This decision, so favourable to the atmospheric system, cannot be wondered at.   The preponderance of evidence, even of engineers, was undoubtedly in its favour; and, however we may now be convinced of the validity of the objections urged against it, and of the superior judgement of the witnesses who opposed it, it was not to be expected that a Committee so composed should be in a position to attach such weight to these objections as to invalidate the concurrent and positive testimony adduced on the other side.

Supported, therefore, by so powerful and public a recommendation, we should be prepared to expect that the atmospheric system would soon have been extensively introduced, at least on new lines—if, indeed, it did not supersede the established plans of locomotion on old ones; for as has been already stated, the inducement to lay down new lines with gradients and works adapted for the plan must have been very strong.

But this official recognition of the merits of the system forms the culminating point of its history; for, strange to say, from this event we have only to trace its continual decadence, and, within but a few years afterwards, to chronicle its abandonment altogether.

The examination of the atmospheric bills in Parliament, in the following session of 1845, proceeded in due course; but the labours of Lord Howick's Committee do not seem to have had the effect intended; for the opponents of the atmospheric system—who, though small in numbers, were very energetic and determined—did not choose to yield to their decision, and the contest had to be renewed in every Committee on every bill.

One of the most important of these contests was that for the line from Newcastle-upon-Tyne to Berwick, in which a single atmospheric line was opposed to the double locomotive railway projected by George Stephenson and his son. The battle of these two rival lines was fought very obstinately and at great length before the Committee of the House of Commons; and in July 1845 they reported that, although they did not feel called upon to express an opinion on the relative merits of the atmospheric and the locomotive systems, or on their comparative applicability to railways in general, still as the evidence negatived the sufficiency of a single line to convey the required traffic, they preferred the locomotive double line, which was eventually adopted.

The Bill for the Epsom and Portsmouth line passed on the atmospheric principle, but was never carried into execution: all the other Bills were either lost or abandoned.

It now only remains to mention the several cases where the atmospheric system has been applied in actual practice; there are five in number.

The first of them, the experimental half mile on the small railway at Wormwood Scrubs, may be dismissed very briefly. It was, as has been stated, set to work in June 1840. It was not intended for traffic, as the line was not commercially used at that time; but it was worked experimentally. During the first year trips were run regularly twice a week, to which the public were admitted free; and subsequently experiments were made at various intervals during $1\frac{1}{2}$ or 2 years more; after which the apparatus was removed.

Next in order comes the Kingstown and Dalkey line, of which a description has already been given. It was first tried in August 1843, and commenced working regularly towards the end of September; but in consequence of some legal difficulties it was not opened for public traffic till some months later. During the interval, however, it was at work, carrying passengers without charge; and any persons who had an introduction, or any special claim, were allowed to make experiments in any manner they desired, without expense to them: the public travelling at the same time for amusement in very large numbers. In March 1844, it began running for regular commercial traffic, and worked for several years, conveying great numbers of passengers to and fro, with perfect safety and considerable speed. On occasions of any peculiar attraction, the double journey was performed every ten minutes, from 7 A.M. to 11 P.M.

About 1855 the Dalkey, Dublin, and Kingstown property was leased to the Dublin, Wicklow, and Wexford Railway Company. The portion between Kingstown and Dalkey was included, and was extended to Bray, where it joined the direct Dublin and Wicklow line; and as it would have been obviously inconvenient to have an isolated portion, of one mile and a half in length, of atmospheric propulsion in the middle of a system worked by loco-motives, the tube was taken up, and the line enlarged, and made a homogeneous part of the extended system.

The third application of the atmospheric principle was upon the London and Croydon Railway. Reference has been made to the Act obtained in 1844 for an extension of that line from Croydon to Epsom, which the Company proposed to work on the atmospheric system. But as the line from Croydon to London was becoming much occupied by the Brighton and the South Eastern Companies, both of which ran over its whole length, it was considered expedient to lay down an additional or third line of rails alongside the other two, over this distance, so as to give an independent accommodation to the Croydon and Epsom traffic. This was considered a favourable portion on which to test the invention; and as this line could be sooner constructed than the new one beyond Croydon, great efforts were made to have the atmospheric apparatus at work upon it as early as possible : and it was opened for a distance of five miles, from Forest Hill to Croydon, in the latter part of 1845. From London to Forest Hill the atmospheric apparatus was not completed, the trains being worked by locomotives, and run into a siding for attachment to the piston carriage of the atmospheric tube.

The tube, which was 15 inches internal diameter, was divided into two lengths—one of 3 miles from Forest Hill to Norwood, and the other of 2 miles from thence to Croydon—there being steam engines of 100 nominal horse-power, with air pumps, placed at each of the three stations. In one place the railway crossed over the Brighton main line by a viaduct, with gradients on each side of 1 in 50.

For a short time after the apparatus was set to work. it was employed in running empty trains at certain intervals during the day, to give the public an opportunity of seeing the new mode of conveyance. It is stated that trains of nineteen carriages were conveyed at 30 to 35

miles an hour, or even at greater speed under favourable
circumstances; but the vacuum employed was high, being
24 to 26 inches, and the power consumed in pumping
was large.* With lighter trains velocities of 60 miles
an hour were attained.

On January 19, 1846, it commenced working for
regular traffic; but, though generally successful, frequent
interruptions took place from various accidents; principally through defects in the stationary engines, which
appeared not to be well adapted to the purpose they had
to serve.

In May the number of trains was increased from thirty-
two to thirty-nine per diem, and the regularity was
generally improved, though some difficulty still occurred
in getting over the steep incline.

The summer discovered an unexpected weakness, for
the June sun, giving a temperature of 131° over the pipe,
acted so powerfully upon the waxy sealing composition
of the valve, that it was difficult to keep it tight against
the pressure of the atmosphere; and on this, as on other
occasions, the aid of the locomotive had to be called in.
The valve itself was also defective, and a new one had to
be inserted, as well as a new sealing material to be com-
pounded, which proved successful, and was suitable to
resist the effect of both heat and cold. The apparatus
was set to work again in July, and a good regularity
attained. A record of three trips on the 21st of that
month † shows satisfactory results. A train of 50 tons

---

* These particulars were given to
the author by Mr. Edward Woods,
who examined and reported on the
apparatus with a view to its adoption in the tunnels at Liverpool,
which, however, he did not recommend.

† Railway Chronicle, 1846, p. 719.

attained a maximum speed of 30 miles an hour, and one of 22 tons, $64\frac{1}{4}$ miles, the vacuum in all cases being 19 inches.

In this year the amalgamation of the Croydon Railway with the London and Brighton line took place. The directors of the former, in giving up their charge, August 26th, remarked that, ' though the atmospheric system had not been free from those difficulties which usually attend the introduction of all new inventions, and though the working expenses had necessarily been very great, still it was progressing satisfactorily.' At the first meeting of the amalgamated companies on August 19, the directors also stated that ' the working progressed satisfactorily, and attained daily a greater degree of regularity, and that there was every reason they should have confidence in it.' Considering, however, the thing still as under experiment, they resolved, on the recommendation of Mr. Cubitt, their engineer, to open the Croydon and Epsom extension, in the first instance, as a locomotive line, until the merits or defects of the new system should have been more thoroughly tried.

In November the manager of the line was directed to make out a statement of the cost of working the system, which he did, much in its disfavour; but his statement was called in question by Mr. Samuda, who contended that the facts did not warrant this disparaging judgment.

During the winter the number of trains was thirty-six daily. In January 1847, a further portion of the atmospheric tube was finished towards London, and on the 14th of that month a trial trip was made, preparatory to the opening. From New Cross to Forest Hill the line ascended for nearly the whole distance an incline of 1 in 100, and the train ran up this and on to Croydon in a satisfactory manner.

In February some stoppages took place in consequence of snow and frost, and the locomotive had again to be resorted to; and at a meeting on the 19th of the same month, the second after the amalgamation, the directors reported that, with a view to determine the amount of expenditure, and at the same time to test practically the value of the system, they had entered into an arrangement with Mr. Samuda for working the atmospheric traction upon a contract for a fixed sum and during a certain period. Some of the shareholders at this meeting strongly advocated the abandonment of the plan; to which it was answered that it had been proved both practicable and efficient, but that the directors would not consent to continue it longer than they thought right.

From this time it seems to have worked well; and a committee of scientific men was appointed by the Institution of Civil Engineers to make experiments on it with a view to determine its powers; when an event occurred which we may best describe by an extract from one of the railway journals of the period, dated May 8, 1847.

Engineering London was suddenly thrown into unusual excitement on Tuesday last by the announcement that the Croydon Atmospheric Pipes were pulled up and the plan abandoned.

On making inquiry we found that it had been decided to abandon the system, that the atmospheric was not in operation, that locomotives were doing the work, and that the atmospheric was doomed.

We confess our surprise at this sudden resolve. The same resolve might have been taken any time these twelve months with more show of reason than appears now on the face of the question. Never before was the atmospheric doing so well,

going so regularly, working so economically. The directors
have for a couple of months been working a contract with Mr.
Samuda, which contract gives them atmospheric power at less
cost than the locomotive; and Mr. Samuda is said to have been
well pleased with his contract and the public service well performed.

We are the more sorry for this resolution, because, although
we have from the beginning been regarded by the advocates of
the atmospheric system as its inveterate enemies, we have really
opposed only what appeared to us the errors of the system; and
while opposing its erroneous application, we have earnestly
supported its having a fair trial. That trial we thought it
would have had on the Croydon, and we are disappointed at this
sudden resolution of the Board, which will, we think, give the
advocates of the system something to complain of, and deprive
all parties of the advantage of an unbiassed decision.

The explanation given by the directors is in their
report of August 10, where they say : 'From the insufficiency of power by atmospheric traction to work the
Epsom in addition to the Croydon traffic, your directors,
by the advice of their consulting engineer, have sub-
stituted locomotive power.' This 'insufficiency' arose
from the vacuum tube being too small. The temptation
to save as much as possible in the first cost of the
apparatus, of which the tube formed such a large item, led
to its being fixed at dimensions which, though probably
large enough for the traffic existing at the time it was
designed, did not allow for much increase. Hence, when
larger loads had to be conveyed, it became necessary
to work the vacuum higher, which, as had long before
been predicted by Mr. Stephenson, brought the elements
of leakage and friction into most disadvantageous operation. Mr. Samuda himself always stated the most eligible
vacuum to be 15 or 16 inches, but with the size of tube

on the Croydon line this vacuum had to be much exceeded when the loads became heavy.

It is possible that this difficulty might have been overcome by dividing the loads, and running trains at more frequent intervals : but there was another motive which probably acted more strongly with the directors than the 'insufficiency of power.' When the atmospheric system was originally adopted by the Croydon Company, their new or third line was to be an isolated one, doing the Croydon traffic only ; but the case was materially altered when this became a trunk line for the Epsom traffic, and for probable future extensions.

Moreover, there had arisen a new management, who had not taken any part in the anterior proceedings. The Croydon Company had sold themselves and their undertaking to a more powerful body, owning a large and important group of lines all worked by locomotive power, under one management and with one stock, except this small piece of atmospheric line, which was so isolated as to be obliged to be connected with locomotive lines at each end.

No doubt, therefore, the Brighton directors were only too glad of any reasonable excuse that might offer for throwing the thing overboard altogether. This excuse came in the sudden pressure of the Epsom race week ; and in spite of the improved behaviour of the apparatus, in spite of the beneficial contract with Mr. Samuda, and in spite of the absurdity of incurring all the cost of the experiment without gaining any intelligible result therefrom, the atmospheric system was forthwith suddenly condemned ; the pipes were taken up and sold for old iron ; the engine houses were pulled down and carted away as old bricks; and thus ended the trial of the atmospheric system on the Croydon Railway.

The next application to be recorded was on the South Devon Railway, a line running between Exeter and Plymouth. The Act incorporating the Company was passed in July 1844; and immediately afterwards a proposal was made by the promoters of the atmospheric system to apply it on that railway. The question was referred to Mr. Brunel, the engineer, who from his examination of its working on the Wormwood Scrubs and the Kingstown lines, came to such a favourable judgment upon it as induced him to recommend it for the line in question, which, having in some parts very difficult gradients and curves, offered a good opportunity for the display of its advantages. His view was confirmed by a committee of the directors, who had been deputed also to examine the working of the system; and it was accordingly resolved to apply it upon the whole line. The railway was laid out expressly for the system, having a single line only, with rails weighing 50 lbs. to the yard, and with bridges and viaducts lighter than those on a locomotive line, and otherwise different in construction.

It was decided to commence the working on the portion of the line between Exeter and Newton—twenty miles with easy gradients. The pipe over this part was 15 inches diameter, and was in six divisions, with a pumping engine at each station. On the steeper and more difficult parts of the line, between Newton and Plymouth, having gradients of 1 in 50 and 1 in 42, it was proposed to have larger pipes, with an expanding piston. The tubes on this line were placed below the level of the rails, to facilitate the formation of level crossings ; and the piston was made to lift up when required.

From the desire to profit as much as possible by the experience acquired on the Croydon Railway, and from other causes, the manufacture of the atmospheric apparatus progressed very slowly; and in the beginning of 1846, a portion of the line being otherwise ready, the engineer decided not to delay the opening any longer, but to commence the passenger traffic with locomotives, which was done from Exeter to Teignmouth on May 30, 1846.

In the beginning of 1847 the stationary engines were erected, but it was April before any length of the atmospheric system was completed, the first trip being satisfactorily made from Exeter to Dawlish, $12\frac{1}{2}$ miles, on the 24th of that month.

In August 1847 it was ready as far as Teignmouth, $15\frac{1}{4}$ miles, and experimental trips were run over it with considerable speed, ease, and precision. With a 30 ton train a speed of 67 miles an hour was obtained; with 50 tons, 60 miles ; with 100 tons, 37 miles. In September the general traffic was worked by it over this distance with apparent satisfaction to the public.   Towards the end of the year it was finished to Newton, and after several successful experimental trials it was publicly opened January 10, dispensing with the locomotives, and running with speed and regularity.

The apparatus, however, appeared liable to some sources of trouble, for at the general meeting in February 1848 Mr. Brunel reported that, notwithstanding numerous difficulties, he thought he was in a fair way of shortly overcoming the mechanical defects, and of bringing the whole

into regular and efficient practical working, so as to be
enabled to test its economy, which its incomplete state
had not till then allowed him to do. At this same meeting
also, the directors announced that, although the atmospheric works were in progress from Newton to Totnes,
a distance of nine miles, comprising difficult ground and
steep inclines, they had decided to delay extending the
system beyond the latter place (excepting only for assistant
power on certain inclines) until experience should have
afforded unquestionable data upon which to estimate its
advantages, and should have confirmed the favourable
opinion which the directors continued to entertain of its
practical efficiency. In the interim it became necessary
to strengthen the works, so as to fit them for locomotive
traffic ; and this being done, the line was gradually finished
from Newton towards Plymouth, and was opened, with
locomotive power, to the immediate vicinity of the latter
place in May 1848.

By this time it was found that the cost of working the
atmospheric system had been much greater than the
directors had reason to anticipate, and, moreover, that
serious defects were beginning to manifest themselves in
the mechanism of the longitudinal valve, the leather of
which was undergoing an unexpected and rapid destruction. These serious considerations led the Board, in July,
to refer the investigation of the whole subject to the
special consideration of a committee, who for many weeks
devoted their attention to it, in constant communication
with the engineer. The result of their labours caused the
directors, at a general Board meeting held August 28,
1848, to pass the following resolution :—

That the very heavy expenses incurred in working on the atmospheric principle between Exeter and Newton, arising in part from the imperfect state and rapid decay of the longitudinal valve, and in part from other causes affecting the system, render it necessary to suspend the employment of it, at the charge of the Company, until the patentees and Mr. Samuda shall have adopted some means, to the satisfaction of the directors, for relieving the Company from the loss consequent upon working under such disadvantages.

It was shown, however, by a document subsequently circulated by the Board, that the defect of the longitudinal valve was not the only difficulty of importance to be overcome. Many others were experienced which weighed greatly on the question of continuing the system :—

1. The necessity for dividing the passenger trains, and reducing the weight of the goods trains, to avoid the chance of all unusually heavy loads.

2. The loss of engine power throughout the line, whenever delays arose in the arrival of the trains ; it being necessary, in the absence of any telegraphic communication, to keep up the vacuum, at enormous cost, until it was required to be used.

3. The other difficulties of working in immediate connection with a main line of near 200 miles, worked upon another system.

4. The probability, if not (under the circumstances of the Company) the certainty, that the atmospheric system could not be adopted on the whole line to Plymouth.

These difficulties, added to the cost of working, and the defective state of the valve, were found so formidable, that the continuance of the atmospheric mode of propulsion, upon an isolated length of twenty miles, connected at each end with lines worked on a different system, became all but impracticable.

In accordance with their resolution, the directors stated, in their report to the general meeting on August 29, that ' Without pronouncing any judgement as to the ultimate success of the atmospheric system, and whilst they are prepared to afford to the patentees and other parties interested in it the use of their machinery for continuing their own experiments, they have agreed with Mr. Brunel that it is expedient for them to suspend the use of the atmospheric system until the same shall be made efficient at the expense of the patentees.'

The operation of the system was accordingly brought to a close on September 9, 1848, and the line thenceforward was worked throughout by locomotives only.

But by far the most complete trial of the atmospheric system has been made in France; and as it does not appear that any account of this experiment has been published, the circumstances may be stated in some detail.*

It appears that the system had at an early period excited some interest in that country, and a French improvement, of much ingenuity, was proposed in its machinery.   This was the invention of M. Hallette, a manufacturing engineer of Arras.   It was a new kind of longitudinal valve for the vacuum main, consisting of two small inflated elastic tubes, fixed in grooves on each side of the longitudinal opening on the top of the pipe, and between which the rod attached to the piston should

---

* For the information contained in this notice we have to express our   obligations to M. Eugène Flachat, the engineer of the line.

slide, the tubes closing again behind it by their own elasticity as it passed along. M. Hallette laid down, at his own expense, an experimental tube, which exhibited his invention in action, and which is said to have worked well; but the ingenious inventor died in 1846, and his project never proceeded farther.

When the Kingstown and Dalkey line was first set to work, the French Government sent M. Mallet, one of the divisional inspectors of the Ponts et Chaussées, to examine its working. His report, which has been translated and published in England, is dated January 10, 1844 : his favourable account of the system appears to have determined the Government to try it in France, and a sum of 1,800,000 francs was accordingly voted for the cost of the necessary apparatus.

The railway on which it was decided to make the trial was that from Paris to St. Germain. This line is altogether about 12½ miles long. At the forest of Vesinet, about 11 miles from Paris, it crosses the Seine, and from thence ascends by a rapid acclivity nearly 170 feet to the plateau on which the town of St. Germain stands. It was on this last 1½ mile that the atmospheric system was applied. The length over which the tube extended was 2,230 metres. For the first 390 metres the line was nearly level; the following 840 metres consisted of a series of inclines, beginning with 1 in 200, and gradually increasing to 1 in 30; and the last 1,000 metres was uniformly 1 in 28¾. On this steep part there was also a curve of 397 metres radius and 400 metres long; two curves on the lower portions were 1,000 metres radius.

The line was double, but the tube was only placed on

the ascending line, the trains running down the other line by their own gravity.

The tube was 63 centimetres (about 24½ inches) internal diameter, calculated for a maximum load of 70 tons. It was of cast iron 2 centimetres thick, strengthened by ribs, and having large feet cast on the lower part to fasten it down. The rails were fixed on longitudinal sleepers, and the tube rested upon the cross transoms which retained the longitudinal timbers in gauge. The longitudinal valve and the other parts of the apparatus were similar to those used in England.

There were two high pressure exhausting engines of 228 horse-power each, which were calculated to cause the ascent of the trains in five or six minutes.

The Company at first proposed to get the apparatus made in France, but were obliged ultimately to have the pipes cast in England. They were put in hand in the beginning of 1846, and the works of the line were ready in the autumn of that year; but by delays in the manufacture of the propelling apparatus the line was not opened for traffic till 1847.

The money voted by government paid for the tube and the engines; the remainder of the outlay, amounting to about 3,200,000 francs, was borne by the Railway Company.

The traffic consisted of passenger trains every hour of the day, for sixteen hours, giving sixteen trains per diem in each direction.

For about six years the average weight of the trains was about 35 tons, and the service was performed with great regularity; but after that time the traffic began to increase, the weight of the trains gradually augmenting to

50 or 60 tons, the consequence of which was the introduction of irregularities in the working. The causes of
these irregularities were well investigated. The principal
one was not chargeable to the system, and might easily
have been remedied—namely, the inadequacy of boiler
power in the stationary engines; but the loads soon began
to approximate closely to the maximum limit of power of
the apparatus, and as it was not always practicable to
determine beforehand the exact weight of the train, it
frequently happened that trains were sent up, on Sundays
and fête days, of a weight touching closely upon this
limit—even although, on arriving at the foot of the steep
incline, the high vacuum of 70 to 72 centimetres (28
inches) was obtained in the tube. The natural result of
this close working was, that if, as occasionally happened,
the train was a little heavier than was calculated, or if
any accidental increase to the resistance arose, the power
proved insufficient for the traction, and the train came to
a stand, or at least did not approach the terminus with
the velocity necessary to shoot it up to the platform after
the pressure had ceased to operate on the piston. The
exhaustion varied usually from 40 centimetres (15½ inches)
to 72 centimetres, according to the weight of the train;
and a singular coincidence was remarked on this line—
that the number of centimetres of exhaustion accurately
denoted the number of tons weight which that degree
of exhaustion would convey. The speed was slow upon
the steep incline, but the trip was performed regularly in
five or six minutes, as intended. Frost and snow were
found to have a prejudicial effect on the valve: the
leather hardened, ice or snow insinuated itself into the

interstices, and the result was increased leakage and extra trouble to keep the machine in efficient action.

After all, however, so long as the haulage power was not overborne by the weight of the trains, the traffic went on pretty well, and the system continued in tolerably successful use for nearly fourteen years—namely, till 1860 — during the whole of which time there had never been a single accident or suspension of the service.

At this time it was found that serious repairs were required to the permanent way, the timber sleepers being decayed. The question then arose whether it might not be preferable to do away with the atmospheric system and to work the incline with locomotives like the rest of the line ; and the following reasons seem to have been considered of sufficient weight to warrant this determination being adopted:—

First, the tractive power was obviously becoming more and more insufficient to work the constantly increasing traffic. Attempts had been made to increase the number of trains to three per hour during the heaviest pressure ; but this led to difficulties at the stations, and it had become necessary on special days to get help from locomotive engines constructed for the purpose. Moreover, in the last year of working, a new element had been introduced, tending still farther to limit the useful power of the apparatus. The carriages had at first been very light, made especially for the purpose ; but it was found desirable to assimilate them to the other stock, and so to make them heavier, which of course, under the limit of weight, diminished the accommodation afforded for passengers by each train. And with this insufficient power, particularly considered in reference to a still farther prospective increase of traffic, there was no chance of any available remedy, if the plan was to be retained. The exhaustion had been carried to its utmost possible extent, and no alternative remained but to lay down a new tube of larger size and engines of larger power, which was clearly out of the question, from its enormous expense and the dead loss of all the expenditure previously incurred.

In the second place, the working expenses had been found very heavy; and although an accurate comparison could not then be made, it was believed that the incline could be worked by locomotives for less expense. The forcing up of the exhaustion, necessary to do the increased work, had augmented disproportionately the consumption of fuel; and as coals had latterly been very dear, the cost of working had showed to great disadvantage.

Then, thirdly, the improvements made in locomotives in late years had removed all doubt as to the practicability of applying them effectively on the steep incline, which could not have been attempted, with much chance of success, when the line was originally laid down.

These arguments appear to have had sufficient weight to lead the directors to abandon the atmospheric system of traction. The tube was accordingly taken up; and the incline is now worked with powerful locomotives constructed expressly for the purpose, and which are said to be able to draw trains of 120 or 130 tons up the incline at less cost than on the former plan.

Such is the history of this remarkable scheme, which, as regards the magnitude of its pretensions, and the interest it excited, has no parallel in railway history. It is scarcely likely to be revived, and therefore it would be useless now to reopen a discussion upon it.   But it may not be out of place, to add a few remarks on the results of the trials made.

We should naturally look to these trials for evidence on three main points—namely, first, the mechanical efficiency of the system as a propelling power ; secondly, its economy ; and thirdly, its general applicability to railway traffic.

With regard to the first head, we can scarcely avoid the conclusion that the trials were, at least, sufficient to establish it as an efficient means of propulsion, considered in a mechanical point of view.

Mr. Stephenson, who was no mean judge in such matters, always testified, with the candour and liberality that distinguished his character, to its mechanical success, and indeed never called its efficiency in question ; and Mr. Bidder declared he considered the mechanical problem as solved beyond doubt.

On the Dalkey line, the system worked the traffic regularly for eleven years.   The Croydon experiment was attended with many vicissitudes, and formed in fact the principal school for the testing and improvement of the machinery on a large scale; but at the time the system was abandoned the mechanical defects had been in a great measure overcome, and it was working more satisfactorily than it had ever done before ; and it is evident that the causes for its discontinuance, on this line, arose more from general policy than from mechanical considerations.

The atmospheric system on the French line worked, while moderately loaded, with great certainty; it was only when it began to be taxed too near the maximum limit of its capability that irregularities occurred; but as, even under all circumstances, it worked for sixteen years without a single accident or suspension of the service, it is clear that no serious objection on mechanical grounds can have appeared.*

On the South Devon line the regularity, speed, and safety were unquestioned. Great prominence was, indeed, given to the defective state of the longitudinal valve, as a reason for its discontinuance; but had this been the only reason, it is difficult to conceive that, under the skill of such an engineer as Mr. Brunel, the same perseverance that had overcome the difficulty on the Dalkey and French lines would not have succeeded on this line also. We have seen, however, that other reasons obtained for the abandonment of the atmospheric system, and there is little doubt that these had more weight in the decision than any mechanical inefficiency.

Great credit is due to the inventors and original engineering promoters of the scheme for the perfection to which it was brought. The original perception of the practicability and advantages of a plan, which, to most minds, would have seemed only a wild vision, was in itself no common merit; and considering the entire

---

* See Perdonnet, Chemins de Fer, vol. ii. chap. xi. page 348, 2nd edit. where M. Flachat recommends the use of the atmospheric pressure for inclined planes. He says: 'Le chemin de fer atmosphérique de St. Germain n'a jamais failli; jamais un accident ne s'est produit; la sécurité du service y est absolue, sa félicité est telle qu'il me semble mériter à ce titre l'attention la plus sérieuse des ingénieurs.' This was in 1860, after fourteen years' trial.

Some English engineers also still retain the opinion that the atmospheric system might be advantageously applied in the present day, in peculiar cases which offer difficulties to the use of the locomotive.

novelty of the whole system and the absence of anything like precedent, the mechanical ingenuity and practical skill exhibited in designing and carrying out the details, was such as to place the contrivers in the highest rank of mechanical engineers, and to elicit the warmest commendation from even the opponents of the plan.

On the question of the *economy* of the system, the evidence is less satisfactory. In almost every instance the working expenses were complained of as very high ; and although the circumstances were in no case such as to render the result absolutely conclusive, we may at any rate consider the question of economy as standing where the arguments of Mr. Stephenson left it ; if not indeed that his opinions were rather confirmed than disproved.

Then, thirdly, as to the *general applicability of the system* to railway traffic—it would seem that the fact of the entire abandonment of the system in every case is, to a certain extent, an argument pointing to a negative conclusion. If the invention had really promised to be beneficial, it is difficult to believe that it would not have been more fully persevered in ; and we can only conceive its abandonment to have been dictated by a strong practical feeling that, even though further perseverance might establish its mechanical and economical success, it would still be found, on other grounds, an ineligible means of locomotion.

It will be seen that Mr. Stephenson's principal objections to the system (apart from the cost) referred to its application to long lines. He urged that for any considerable length of railway, a double line with a complete double apparatus was absolutely essential; and that even with this, and though the economy were in its favour, yet on railways of large extent, there must exist conditions which would militate against its certainty of action, and which must disqualify it for being an appropriate means of railway traffic. Now it is quite clear that none of the trials actually made were of a nature to touch these objections. The longest line tried—the South Devon—had none of the characteristics of traffic on large trunk lines to which Mr. Stephenson's reasonings applied, and therefore we must consider that his arguments on this head remain in full force, notwithstanding anything that has been done.

The immediate cause of the abandonment of the system, sooner or later, in every case where it has been tried, appears to have lain in its inflexibility—its want of elasticity—in its incapability of adapting itself to the changeable requirements and circumstances of a variable traffic—in its very peculiar nature, so uncongenial to the established habits of railway people—and in the great difficulty of bringing it to work conveniently and harmoniously in conjunction with other systems of railway traction. If we could conceive a line of railway isolated from all others, and where the traffic should be perfectly uniform in amount and regular in time, possibly, as Mr. Stephenson admitted, the atmospheric system might be there applicable with advantage; but such a line would be an exceptional one; and certainly none of the railways on which it has been tried have approximated to these conditions.

An examination will show that, in every case, the most urgent reasons for the abandonment of the plan lay, either in the increase of traffic beyond what the tube could work, or in its isolated condition between locomotive lines at each end, which rendered the break of the system of haulage peculiarly disadvantageous, and fraught with such inconveniences as the proprietors would not submit to. On the French line the former of these objections prevailed; on the Dalkey and South Devon the latter; on the Croydon line both combined.

The inflexible and unaccommodating nature of the system was often and strongly insisted upon by Mr. Stephenson as a most powerful objection to it, applying indeed to every system of haulage by stationary power. It had been prominent in the original discussions on this subject in 1830, and it was obvious that the atmospheric system was only a renewal of the old proposition in a new form.

The system aimed at too great a change. It was not a mere improvement in things already existing — it was an entire revolution; a total subversion of the established mode of conducting the traffic, and a substitution of an entirely new plan: we cannot therefore wonder that it met with great opposition; nor could it be expected that anything short of the most complete and triumphant superiority could establish it. Railway people had become attached to the locomotive from its extreme convenience; and the change to a more rigid and limited plan was certainly not likely to find favour. There may be something in the national English independence of character which led railway officials to prefer a system that they could manage and vary with full liberty, to one in which they would all become, as it were, mere parts of one huge machine.

For railways, generally, the locomotive appears now too well established to be liable to farther opposition from any modification of stationary power. It is true that it is, and must ever be, subject to many disadvantages inherent in the travelling form of the machine; but, considering the great improvements which have been made in it of late years, and its modern success in cases where its application was long considered impossible —and taking into account its versatile adaptability to variations of traffic ; its admirable suitability to sudden emergency ; and its wonderful convenience of management and control—we think there can now be little dissent from the opinion so resolutely maintained by Mr. Stephenson, that the system of traction which rendered his father's name famous is the only one well fitted for general use upon railways.

W. P.

It may be useful to put on record the following list of published authorities made use of in this chapter : —

Acta Eruditorum. Leipsic 1688.

A New Method of conveying Letters and Goods with great Certainty and Rapidity by Air. By G. Medhurst, Inventor, Patentee, and Proprietor, 1 Denmark Street, Soho. London 1810.

Calculations and Remarks tending to prove the Practicability, Effects, and Advantages of a Plan for the rapid Conveyance of Goods and Passengers upon an Iron Road, through a Tube of Thirty Feet in Area, by the Power and Velocity of Air. By G. Medhurst, Inventor and Patentee, Denmark Street, Soho. London 1812.

On Facility of Intercourse. By John Vallance of Brighton. London 1824.

A New System of Inland Conveyance for Goods and Passengers, capable of being applied and extended throughout the Country, and of Conveying all kinds of Goods, Cattle, and Passengers, with the Velocity of Sixty Miles in an Hour, at an Expense that will not exceed the One-fourth Part of the Present Mode of Travelling, without the Aid of Horses or any Animal Power. By George Medhurst, Civil Engineer, Denmark Street, Soho. London 1827.

A Treatise on the Steam Engine, Historical, Practical, and Descriptive. By John Farey, Engineer. London 1827.

Report to the Directors of the Liverpool and Manchester Railway on the Comparative Merits of Locomotive and Fixed Engines as a Moving Power. By James Walker and J. U. Rastrick, Esq., Civil Engineers. Liverpool 1829.

Observations on the Comparative Merits of Locomotive and Fixed Engines, as applied to Railways: being a Reply to the Report of Mr. James Walker to the Directors of the Liverpool and Manchester Railway, compiled from the Reports of Mr. George Stephenson. With an Account of the Competition of Locomotive Engines at Rainhill in October 1829, and of the subsequent Experiments. By Robert Stephenson and Joseph Locke, Civil Engineers. Liverpool 1830.

Clegg's Patent Atmospheric Railway. London 1839.

Clegg and Samuda's Atmospheric Railway. London 1840.

Irish Railways. The Atmospheric Railway. A Letter to the Rt. Hon. Lord Viscount Morpeth. By James Pim, jun., Treasurer of the Dublin and Kingstown Railway Company. London 1841.

The Atmospheric Railway. A Letter to the Rt. Hon. the Earl of Ripon, President of the Board of Trade, &c. &c. By James Pim, M.R.I.A. Treasurer of the Dublin and Kingstown Railway Company. With Plates. London 1841.

A Treatise on the Adaptation of Atmospheric Pressure to the Purposes of Locomotion on Railways. With Two Plates. By J. D'A. Samuda, London. (The date on the title-page is 1844, but the real date of the pamphlet is 1841.)

Report of Lieut.-Colonel Sir Frederic Smith, R.E., and Professor Barlow to the Rt. Hon. the Earl of Ripon, President of the Board of Trade, on the Atmospheric Railway. Presented to both Houses of Parliament by command of Her Majesty. London 1842.

Report on the Railway constructed from Kingstown to Dalkey in Ireland, upon the Atmospheric System, and upon the Application of this System to Railroads in general. By C. Mallet. Dated Paris, January 10, 1844.

Report on the Atmospheric Railway System. By Robert Stephenson, Esq. London 1844.

Croydon and Epsom Railway, &c. &c. Minutes of the Evidence of the Engineers examined before the Committee on the Croydon and Epsom and South Western and Epsom Railway Bills, with reference to the Working of Railways upon the Atmospheric Principle. Ordered by the House of Commons to be printed June 10, 1844.

Report from the Select Committee on Atmospheric Railways; together with the Minutes of Evidence. Ordered by the House of Commons to be printed April 24, 1845.

Minutes of Proceedings of the Institution of Civil Engineers. London 1844 and 1845.

Railways: their Rise, Progress, and Construction, &c. By Robert Ritchie. London 1846.

Tube Propulseur, Hallette, &c. &c.   Paris.
The Railway Chronicle.   London 1846 to 1848.
Reports of the South Devon Railway.
Traité Elémentaire des Chemins de Fer. Par Aug. Perdonnet.   Paris 1860.

END OF THE FIRST VOLUME.

LONDON
PRINTED BY SPOTTISWOODE AND CO.
NEW-STREET SQUARE

Printed in the United States
By Bookmasters